Lecture Notes in Physics

Springer-Verlag Berlin Heidelberg GmbH

Physics and Astronomy ONLINE LIBRARY

http://www.springer.de/phys/

The Editorial Policy for Proceedings

The series Lecture Notes in Physics reports new developments in physical research and teaching – quickly, informally, and at a high level. The proceedings to be considered for publication in this series should be limited to only a few areas of research, and these should be closely related to each other. The contributions should be of a high standard and should avoid lengthy redraftings of papers already published or about to be published elsewhere. As a whole, the proceedings should aim for a balanced presentation of the theme of the conference including a description of the techniques used and enough motivation for a broad readership. It should not be assumed that the published proceedings must reflect the conference in its entirety. (A listing or abstracts of papers presented at the meeting but not included in the proceedings could be added as an appendix.)

When applying for publication in the series Lecture Notes in Physics the volume's editor(s) should submit sufficient material to enable the series editors and their referees to make a fairly accurate evaluation (e.g. a complete list of speakers and titles of papers to be presented and abstracts). If, based on this information, the proceedings are (tentatively) accepted, the volume's editor(s), whose name(s) will appear on the title pages, should select the papers suitable for publication and have them refereed (as for a journal) when appropriate. As a rule discussions will not be accepted. The series editors and Springer-Verlag will normally not interfere with the detailed editing except in fairly obvious cases or on technical matters.

Final acceptance is expressed by the series editor in charge, in consultation with Springer-Verlag only after receiving the complete manuscript. It might help to send a copy of the authors' manuscripts in advance to the editor in charge to discuss possible revisions with him. As a general rule, the series editor will confirm his tentative acceptance if the final manuscript corresponds to the original concept discussed, if the quality of the contribution meets the requirements of the series, and if the final size of the manuscript does not greatly exceed the number of pages originally agreed upon. The manuscript should be forwarded to Springer-Verlag shortly after the meeting. In cases of extreme delay (more than six months after the conference) the series editors will check once more the timeliness of the papers. Therefore, the volume's editor(s) should establish strict deadlines, or collect the articles during the conference and have them revised on the spot. If a delay is unavoidable, one should encourage the authors to update their contributions if appropriate. The editors of proceedings are strongly advised to inform contributors about these points at an early stage.

The final manuscript should contain a table of contents and an informative introduction accessible also to readers not particularly familiar with the topic of the conference. The contributions should be in English. The volume's editor(s) should check the contributions for the correct use of language. At Springer-Verlag only the prefaces will be checked by a copy-editor for language and style. Grave linguistic or technical shortcomings may lead to the rejection of contributions by the series editors. A conference report should not exceed a total of 500 pages. Keeping the size within this bound should be achieved by a stricter selection of articles and not by imposing an upper limit to the length of the individual papers. Editors receive jointly 30 complimentary copies of their book. They are entitled to purchase further copies of their book at a reduced rate. As a rule no reprints of individual contributions can be supplied. No royalty is paid on Lecture Notes in Physics volumes. Commitment to publish is made by letter of interest rather than by signing a formal contract. Springer-Verlag secures the copyright for each volume.

The Production Process

The books are hardbound, and the publisher will select quality paper appropriate to the needs of the author(s). Publication time is about ten weeks. More than twenty years of experience guarantee authors the best possible service. To reach the goal of rapid publication at a low price the technique of photographic reproduction from a camera-ready manuscript was chosen. This process shifts the main responsibility for the technical quality considerably from the publisher to the authors. We therefore urge all authors and editors of proceedings to observe very carefully the essentials for the preparation of camera-ready manuscripts, which we will supply on request. This applies especially to the quality of figures and halftones submitted for publication. In addition, it might be useful to look at some of the volumes already published. As a special service, we offer free of charge LaTeX and TeX macro packages to format the text according to Springer-Verlag's quality requirements. We strongly recommend that you make use of this offer, since the result will be a book of considerably improved technical quality. To avoid mistakes and time-consuming correspondence during the production period the conference editors should request special instructions from the publisher well before the beginning of the conference. Manuscripts not meeting the technical standard of the series will have to be returned for improvement.

For further information please contact Springer-Verlag, Physics Editorial Department II, Tiergartenstrasse 17, D-69121 Heidelberg, Germany

Series homepage – http://www.springer.de/phys/books/lnpp

J.-P. Rozelot L. Klein J.-C. Vial (Eds.)

Transport and Energy Conversion in the Heliosphere

Lectures Given at the CNRS Summer School
on Solar Astrophysics, Oleron, France,
25-29 May 1998

 Springer

Editors

J.-P. Rozelot
OCA/CERGA
Av. Copernic
06130 Grasse, France

L. Klein
Observatoire de Paris Meudon
DASOP
92135 Meudon, France

J.-C. Vial
IAS, Université Paris XI, CNRS
91405 Orsay, France

Library of Congress Cataloging-in-Publication Data applied for.

Die Deutsche Bibliothek - CIP-Einheitsaufnahme

Transport and energy conversion in the heliosphere : lectures given at
the CNRS Summer School on Solar Astrophysics, Oleron, France, 25 - 29
May 1998 / J.-P. Rozelot ... (ed.). - Berlin ; Heidelberg ; New York ;
Barcelona ; Hong Kong ; London ; Milan ; Paris ; Singapore ; Tokyo :
Springer, 2000
  (Lecture notes in physics ; Vol. 553)
  (Physics and astronomy online library)

ISSN 0075-8450
ISBN 978-3-662-14265-3   ISBN 978-3-540-45166-2 (eBook)
DOI 10.1007/978-3-540-45166-2

© Springer-Verlag Berlin Heidelberg 2000
Originally published by Springer-Verlag Berlin Heidelberg New York in 2000.
Softcover reprint of the hardcover 1st edition 2000

Typesetting: Camera-ready by the authors/editors
Cover design: *design & production*, Heidelberg

Printed on acid-free paper
SPIN: 10771051   55/3141/du - 5 4 3 2 1 0

Foreword

The book contains courses taught to a public of Ph.D. students, post-docs and confirmed researchers in all fields of heliospheric plasma physics. It aims at identifying physical issues which are common to two different fields of astronomy: solar and magnetospheric physics. Emphasis is given to basic processes of transport and conversion of energy: magnetic reconnection is discussed in detail from the viewpoints of MHD and kinetic physics. Processes of charged particle acceleration are reviewed and confronted with recent observations. The subject is introduced by a summary of MHD and the basic structures and parameters of the solar atmosphere, terrestrial ionosphere and magnetosphere are reviewed. The book combines a pedagogic and comprehensive presentation of physical issues and raises fully open questions, with the complementary and sometimes conflicting views of geophysicists and solar physicists. The book's focus, while basic, opens new avenues.

Observatory of Meudon, France	*Ludwig Klein*
IAS, Orsay, France	*Jean-Claude Vial*
OCA, France	*Jean-Pierre Rozelot*
August 2000	*The Editors*

Preface

Following the great success of the first two CNRS Summer Schools on Solar Astrophysics held in Oléron (May 1996 and May 1997 – two schools devoted to the highlights of solar physics), I came to the conclusion that the initiative should be continued. A new programme committee, consisting of

Claudio Chiuderi (Observatory of Arcetri, Florence),
Ludwig Klein (Observatory of Meudon, France),
Joseph Lemaire (Space Laboratory for Astronomy, Brusselles, Belgium),
Alain Roux (CETP, Velizy, France),
Jean-Claude Vial (IAS, Orsay, France),
and *Jean-Pierre Rozelot* (OCA, France),

was set up and embarked on the organization of this third school. The basic idea was that, in view of the recent exciting developments in the physics of the solar corona, the magnetosphere, and boundary between the two, the solar wind, the committee decided it was important to unite the solar physics and magnetospheric communities, which had been separated for so long. Within this framework, people could effectively interact, exchange new scientific ideas and establish a common language to overcome the technical barriers between the two fields. At the same time they could try by analyzing recent breakthroughs, and so make progress in the two major areas of:

- magnetic reconnection;
- and the process of acceleration.

We wished to establish a general framework in which these highly specialized, but intersciplinary fields could be furthered. It seemed to us essential that young researchers should be aware of underlying hypotheses and to acquire the necessary techniques and tools for calculation. Also equally important is that everyone understands the limits that are imposed by the instruments used for understanding the physical environments of the corona and magnetosphere.

This Workshop was thus devoted to *"Magnetic Reconnection"* and the courses covered:

- a general introduction (to MHD), recalling basic physics notions and establishing a common basis;

- a review of classical aspects of magnetic reconnection, which is an important mechanism that changes the toplogy of sheared magnetic fields and converts magnetic energy to both thermal energy and the acceleration of plasma;
- an overview of the role of the magnetic field in the solar atmosphere and how magnetic reconnection can be driven by the emergence of sheared magnetic field;
- a comprehensive introduction to the physics of magnetospheric plasmas, proposing the key approaches needed to gain new insight into the structure and some specific processes the magnetosphere;
- a new self consistent kinetic approach of collissionless plasmas.

In its "Lecture Notes", Springer Verlag gives an excellent opportunity to voice the best of the scientific thought in this field. I am particularly grateful to all speakers, namely, *Giorgio Einaudi, Pascal Démoulin, Olivier Le Contel, Philippe Louarn, Clare Parnell and René Pellat,* for their formal or informal contributions. Moreover, the innumerable individual discussions appeared to fit well with the overall aim of the School. Nearly all the plenary lectures are collected in this volume. From that point, I wish to warmly thank the authors for the careful preparation of their manuscripts.

This monograph, one of the first comprehensive reviews available on the subject is intended for astrophysicists who are seeking an introduction to the physics of magnetic reconnection inside the heliosphere, and for students at graduate level.

Finally, this school would not have been possible without the financial support of the *"Formation Permanente du CNRS"*. This same institution also kindly sponsored the fourth CNRS Summer School, which was held from the 21st to the 25th of June, 1999, in Oléron on the *Data Analysis for Astrophysics and Geophysics*.

Grasse, France
July 2000   *Jean-Pierre Rozelot*

Contents

Introduction to MHD

Jean Heyvaerts

Observatoire de Strasbourg, 11, rue de l'Université, 67000 Strasbourg; *email: heyvaert@astro.u-strasbg.fr*

Part I: Straightforward MHD

1 What Is a Plasma?

A plasma is a conducting fluid, gas or liquid. The plasma state is found in nature under a wide variety of different physical conditions, very dense or very tenuous, very hot or cold. Ionization, total or partial, allows for electric current conduction. It may be caused by the thermodynamic state of the medium, as for example the high temperature of the solar coronal gas, or, as is the case in the interior of white dwarf stars, the high density of the medium, electrons becoming quantum-degenerate. Alternatively, a cold medium can be ionized by some external agent, such as solar UV irradiation in the upper terrestrial atmosphere (the ionosphere) or by cosmic rays penetrating molecular clouds in the interstellar medium. Such a fluid, especially when under-dense, may not be "hydrodynamical" in character, like air or water. We come back on this in some detail in Part II. As for now, we shall simply assume uncritically that a plasma really behaves like a regular fluid and is well described by "Magneto-Hydro-Dynamic" fields (MHD fields for short), which are local macroscopic properties of the fluid which depend on position in space, r, and on time, t, namely:

- The mass density of the fluid $\rho(r, t)$
- Its local velocity $v(r, t)$
- Its local pressure $P(r, t)$
- Its local temperature $T(r, t)$
- The local magnetic field $B(r, t)$
- The local electric current density $j(r, t)$
- The local electric field $E(r, t)$

and, more generally, any quantity necessary for specifying the state of the plasma and of its electromagnetic environment. Some other quantities, as for example the internal energy density, can be expressed in terms of the above ones, which themselves may in some instances not be independent of eachother. The MHD description of plasma evolution rests on the equations obeyed by these MHD fields, which we establish from a fluid dynamics point of view in Part I. This viewpoint however does not exhaust all the physics involved in plasma dynamics. We come back in Part II on this from a gas kinetic point of view. The relations between these two approaches will then be discussed.

2   Conservation Equations

2.1   Densities

In a continuous medium, the distribution in space of extensive quantities, such as mass, energy, electric charge or momentum is characterized by volume-densities, which are functions of position and time.

2.2   Fluxes

A flux characterizes the transport in space of these quantities. The flux of a scalar quantity, G say, is a vector $\boldsymbol{\Phi}$ defined as follows. Let $d\boldsymbol{S}$ be an (oriented) surface element about the point $\boldsymbol{r}$. The amount dG of quantity G that goes through $d\boldsymbol{S}$ between times t and $t + dt$ is expressed as:

$$dG = (\boldsymbol{\Phi} \cdot d\boldsymbol{S})\ dt \tag{1}$$

This relation defines the flux vector $\boldsymbol{\Phi}(\boldsymbol{r}, t)$ at point $\boldsymbol{r}$ and time t. Note that dG is algebraic. It is positive when transport takes place in the sense of $d\boldsymbol{S}$.

2.3   Tensorial fluxes of Vectorial Quantities

If the extensive quantity G is vectorial in nature, such as momentum, its density is also vectorial while its flux is a second-order tensor. Let us precise what a tensorial flux is. Suppose we describe the transport of quantity $\boldsymbol{G}$ in some arbitrarily chosen cartesian rest-frame, $\mathcal{R}$. $\boldsymbol{G}$ has three components, G_x , G_y and G_z which are "numerical" quantities, though not "scalar" ones because their values would change if we were to change reference frame. There is no difficulty, in rest frame $\mathcal{R}$, in defining a density and a flux vector for each of these "numerical" quantities. Let us call them g^b and $\boldsymbol{\Phi}^b$ respectively, the index b taking the values x, y or z. Each of the three vectors $\boldsymbol{\Phi}^b$ has three components which can be labeled by an index a. The a-component of $\boldsymbol{\Phi}^b$ is noted Φ^{ab}. Since both a and b can take three values, there are nine numbers Φ^{ab}. They satisfactorily characterize the transport of the extensive vector quantity $\boldsymbol{G}$, but, at this point, they do so in rest frame $\mathcal{R}$ only. What if we change reference frame? Let us go to another cartesian frame $\mathcal{R}'$. The change from $\mathcal{R}$ to $\mathcal{R}'$ is characterized by a matrix $X_b^{b'}$ which relates the components of a vector in $\mathcal{R}'$, call them $V^{b'}$, to those which it owes in frame $\mathcal{R}$, call them V^b, by:

$$V^{b'} = \sum_b X_b^{b'} V^b \equiv X_b^{b'} V^b \tag{2}$$

In the second term of this expression the very convenient "dummy index rule" has been used. It states that whenever in an expression some index is twice repeated, there is an implicit summation on the values that it may take. For example, in the last term of equations (2), index b is repeated, which implies, according to this rule, that it really means the sum written in the second term of the equation. This transformation rule applies to the components of G, therefore:

$$G^{b'} = X_b^{b'} G^b \tag{3}$$

Moreover, the "flux of a sum" is obviously the "sum of the fluxes" (this is implicit in the definition of a flux), a property which holds more generally true for a linear combination. Therefore the flux of the b' component of G is

$$\Phi^{b'} = X_b^{b'} \Phi^b \tag{4}$$

The quantity $\Phi^{a'b'}$ is the a'-component in rest frame $\mathcal{R}'$ of vector $\Phi^{b'}$. It is obtained from the components $\Phi^{ab'}$ of $\Phi^{b'}$ in $\mathcal{R}$ by the transformation rule of components:

$$\Phi^{a'b'} = X_a^{a'} \Phi^{ab'} \tag{5}$$

The a-component of equation (4) gives:

$$\Phi^{ab'} = X_b^{b'} \Phi^{ab} \tag{6}$$

whence, by (5),

$$\Phi^{a'b'} = X_a^{a'} X_b^{b'} \Phi^{ab} \tag{7}$$

The similarity of this frame transformation rule with the one which applies for vectors is noteworthy. The difference rests in the number of indices needed to label the components. A vector component is labeled by one index while the quantities Φ^{ab} or $\Phi^{a'b'}$ are labeled by two and this reflects in the number of frame transformation matrices needed to pass these quantities from one to the other frame: one matrix for vector components, two matrices for the Φ^{ab}'s. This leads to the idea that a genuinly geometrical object is hidden behind the Φ^{ab} numbers, which are to be regarded as the "components" in $\mathcal{R}$ of this truely geometrical object. One should really think of it as a geometrical entity, independent of the rest frame, eventhough, unlike vectors, a graphical representation of it is difficult. This object is a "second-rank tensor". The notation for it will be $\overline{\overline{\Phi}}$ or sometimes just Φ when there is no ambiguity about its rank. The quantities Φ^{ab} are the "components" of the tensor $\overline{\overline{\Phi}}$ in $\mathcal{R}$ and the quantities $\Phi^{a'b'}$ are the components in $\mathcal{R}'$ of this very same tensor. For simplicity we have defined here only cartesian components of second rank tensors to avoid entering in complications associated with general coordinate systems and distinguishing between contravariant and covariant components. More can be read on this in books dealing with riemannian geometry. There is usually an introductory chapter on this subject in any general relativity book. Operations on tensors are simple (see, for example, in this series, Heyvaerts 1991) and will be introduced here when necessary.

2.4   Volumic Rate of Creation

The quantity G, be it scalar or vector, can usually be created or destroyed. Such processes can be characterized by a volumic rate of creation, s, a scalar or a vector according to the nature of G. This rate is defined such that the amount dG_{cr} of quantity G which appears in an infinitesimal volume d^3r near point r in dt seconds could be written as:

$$dG_{cr} = s\, d^3r\, dt \tag{8}$$

Disappearance of some amount of G is accounted for by a negative value of the creation rate s.

2.5   Conservation Equations

It is a rather trivial statement that the change dG between times t and $t+dt$ of the quantity G present in a small volume $V \equiv d^3r$ can be expressed as the sum of two parts. One is the amount of G that has entered V or has left it through its border B, which we note $dG_{in/out}$ and the other is due to internal changes of the amount of G inside V itself, dG_{inside}, say.

$$dG = dG_{in/out} + dG_{inside} \tag{9}$$

By the definition of the volumic rate of creation s the latter is:

$$dG_{inside} = \left(\int_V s(r,t)d^3r \right) dt \approx sV dt \tag{10}$$

and, by the definition of flux, the former, which is the balance of imports and exports through B, is: written as:

$$dG_{in/out} = \left(\int_{B(V)} \boldsymbol{\Phi} \cdot d\boldsymbol{S}_{in} \right) dt \tag{11}$$

In this expression $d\boldsymbol{S}_{in}$, a surface element normal to the boundary B, is oriented towards its inside, so that imports are counted as positive. Paying due attention to this in using the flux-divergence theorem, we obtain, for infinitesimal V:

$$dG_{in/out} \approx -(\mathrm{div}\boldsymbol{\Phi})\, V\, dt \tag{12}$$

The quantity dG can also be expressed in terms of the time variation of the density g by:

$$dG = d\left(\int_V g\, d^3r \right) \approx V\, \frac{\partial g}{\partial t}\, dt \tag{13}$$

From (9), (10), (12) (13) we obtain the so-called conservation equation for G:

$$\frac{\partial g}{\partial t} + \mathrm{div}(\boldsymbol{\Phi}) = s \tag{14}$$

If the quantity is vectorial, so are its density and volumic rate of creation. Its flux is a second rank tensor $\overline{\overline{\Phi}}$, the divergence of which is a vector with cartesian component-a defined by

$$(\mathrm{div}\overline{\overline{\Phi}})^a = \frac{\partial \Phi^{ba}}{\partial x_b} \equiv \nabla_b \Phi^{ba} \tag{15}$$

The conservation equation for G is:

$$\frac{\partial g}{\partial t} + \mathrm{div}\overline{\overline{\Phi}} = s \tag{16}$$

Exercise: write down the conservation equation for an extensive quantity $\mathbf{G}$ which is second rank tensor in nature; define its flux and make explicit the rules for changing its components when changing reference frame.

3   Derivative Following the Motion

It sometimes turns out to be useful, when dealing with fluids, to apply known laws of physics, such as the fundamental law of mechanics or laws of thermodynamics, to some well defined parcel of fluid. Such a fluid parcel moves and changes shape as it does. An observer riding it observes time-dependent local physical quantities such as density, temperature, average velocity of the parcel. Consider again an extensive quantity G, such as mass or energy or momentum. In the ususal "eulerian" description of the fluid, introduced above, the distribution of G in space is described by its density field, the function $g(\mathbf{r},t)$, where $\mathbf{r}$ is some fixed point in the rest frame used to describe the fluid motion. A particular fluid parcel would have in this frame a motion described by $\mathbf{r}(t)$, say. The density function $g(t)$ seen by an observer which follows it is the "lagrangean" variation of the density (i.e. following the motion of that particular parcel), or lagrangean density for short. It is related to the eulerian density field $g(\mathbf{r},t)$ by

$$g_{lagr}(t) \equiv g(t) = g(\mathbf{r}(t), t) \tag{17}$$

The derivative with respect to time of $g_{lagr}(t)$, also named "derivative following the motion" or "lagrangean derivative" is easily calculated from the derivative chain rule:

$$\frac{dg}{dt} = \frac{d}{dt}\left(g(x(t), y(t), z(t), t)\right) = \frac{\partial g}{\partial t} + \frac{\partial g}{\partial x}\frac{dx}{dt} + \frac{\partial g}{\partial y}\frac{dy}{dt} + \frac{\partial g}{\partial z}\frac{dz}{dt} \tag{18}$$

Note that dx/dt etc.. are the components of the velocity of the parcel of fluid, which allows to condensate this expression as:

$$\frac{dg}{dt} = \frac{\partial g}{\partial t} + \mathbf{v} \cdot \boldsymbol{\nabla} g \tag{19}$$

The lagrangean time derivative should not be mistaken for the eulerian partial derivative with respect to time, $\partial g/\partial t$, which is calculated at a fixed point in the "laboratory" reference frame. This important difference is taken care of by using a specific notation for the lagrangean derivative with respect to time, usually D/Dt, sometimes simply d/dt:

$$D/Dt = \partial/\partial t + \boldsymbol{v} \cdot \boldsymbol{\nabla} \qquad (20)$$

If the quantity under study is vectorial, the definition of the lagrangean derivative of each of its cartesian components remains as given by equation (19): the $(\boldsymbol{v} \cdot \boldsymbol{\nabla})$ operator acting on a vector field $\boldsymbol{A}(\boldsymbol{r}, t)$ gives a vector whose cartesian a-component is:

$$((\boldsymbol{v} \cdot \boldsymbol{\nabla})\boldsymbol{A})^a \equiv (v^j \nabla_j) A^a \qquad (21)$$

4   Mass Conservation Equation

Let ρ be the mass density. The mass flux reduces to its "convected" part $\rho \boldsymbol{v}$. This this is best seen by coming back to the definition of flux and calculating how much mass passes in dt seconds through a surface element $d\boldsymbol{S}$. In non-relativistic physics mass is globally conserved, being neither created nor destroyed, eventhough the mass of particular constituents may change, for example in chemical reactions. So the volumic rate of creation of mass as a whole is nil and the equation of conservation for mass can be written as:

$$\frac{\partial \rho}{\partial t} + \mathrm{div}(\rho \boldsymbol{v}) = 0 \qquad (22)$$

Exercise: Write down the equation of conservation for the number of molecules of a certain particular constituant of the fluid, taking into account chemical reactions represented by an appropriate rate. For simplicity neglect element diffusion with respect to the bulk of the fluid.

5   Electric Charge Conservation Equation

Let ρ_e be the electric charge density. By its definition, the electric current density $\boldsymbol{j}$ is the charge flux. It consists of two parts. One, the so-called "convected current", is due to the bulk motion of electric charge with the fluid. It can be written as $\rho_e \boldsymbol{v}$. The other, the so-called conducted current $\boldsymbol{j}_{cond}$, takes care of the fact that charge is carried by different charge carriers, i.e. electrons, different types of ions etc.., Each species, α, has its own species-velocity $\boldsymbol{v}_\alpha(\boldsymbol{r}, t)$. The velocity $\boldsymbol{v}(\boldsymbol{r}, t)$ of the fluid as a whole is the average value of the $\boldsymbol{v}_\alpha$'s weighted by their mass density ρ_α while the total electric current density is its average weighted by their charge density $\rho_{e,\alpha}$. Since the weighting is different for mass and charge densities, and the $\boldsymbol{v}_\alpha$'s differ slightly from the local global fluid velocity v, by a difference δv_α say, the total current does not simply reduce to its convected part $\rho_e v$. The difference is the conduction current:

$$j_{cond} = \sum_\alpha \rho_{e\alpha}\, \delta v_\alpha \qquad (23)$$

Had we made the same reasoning for mass, we would have found that the "conduction" flux of mass is identically zero, because the fluid velocity as a whole is defined as the average value weighted by mass density of the different species. Summing up, we have shown that the electric current is the sum of two terms

$$j = \rho_e v + j_{cond} \qquad (24)$$

Electric charge being globally conserved (no net charge creation or destruction in any process whatsoever), its volumic rate of creation is nil. So the charge conservation equation is just

$$\frac{\partial \rho_e}{\partial t} + \mathrm{div} j = 0 \qquad (25)$$

This equation results from Maxwell's equation as well, or is expressed implicitly by them if one prefers to think of it that way. Conservation equations, like those for mass or charge, which are devoid of "source term" (i.e. of volumic rate of creation term) are said to be "perfect" conservation laws. This is because the change of the associated extensive quantity in some volume V, infinitisemal or not, is only due to incomes or outcomes through the volume boundary B, implying that the change inside V is compensated by an opposite change in its outside, the quantity being globally exactly conserved when summing inside and outside.

6   Momentum Conservation Equation

6.1   Matter Momentum Conservation Equation

Obviously the density of momentum of matter is $\varpi = \rho v$. What are its flux and volumic rate of creation? To determine them, let us come back to the basics and apply the fundamental law of mechanics to a fluid parcel, following it in its motion. Let $V(t)$ be its (infinitesimal) volume at time t. Though its volume and its density separately change, its mass $m = \rho V$ remains unchanged because it consists of some well defined bit of matter (set of molecules). The acceleration γ of this little body is the lagrangean derivative of its velocity:

$$\gamma = \partial v/\partial t + (v \cdot \nabla)v \qquad (26)$$

The forces exerted on the parcel of fluid are either volume-forces or contact-forces, the latter being due to the action on the parcel of its fluid environment.

Volume forces are characterized by force densities, $\boldsymbol{f}$, such as the gravitational force density

$$\boldsymbol{f}_{grav} = \rho \boldsymbol{g} \tag{27}$$

or the electromagnetic force density

$$\boldsymbol{f}_{em} = \rho_e \boldsymbol{E} + \boldsymbol{j} \times \boldsymbol{B} \tag{28}$$

Other volume forces might be considered, depending on the situation, for example forces resulting from the emission or absorption of radiation by the fluid, which is electro-magnetic in nature, though maybe quantum, and therefore often not macroscopic enough to be conveniently represented in the form of an electromagnetic force density. Let us loosely call these other force densities $\boldsymbol{f}_{other}$. Contact forces can be expressed as integrals on the surface bordering V, $B(V)$. They consist of pressure and viscosity forces. The total pressure force on the parcel is

$$\boldsymbol{F}_{press} = \int_{B(V)} P \, d\boldsymbol{S}_{in} \tag{29}$$

where P is the pressure and $d\boldsymbol{S}_{in}$ the surface element on $B(V)$ oriented to the inside of V, so that the force exerted by the outside onto the inside be obtained. The viscosity forces exerted on an interface element dS are represented by a second rank tensor of viscous stresses, $\overline{\overline{\sigma}}$, such that the viscosity force exerted through dS by the outside on the inside is expressed as:

$$d\boldsymbol{F}_{visc} = \overline{\overline{\sigma}} \cdot d\boldsymbol{S}_{out} \tag{30}$$

The sign convention for the definition of the viscous stress tensor is such that, unlike for the pressure force, the surface element should appear here as oriented outwards. In Part II we shall give an explicit expression for this viscous stress tensor. The dot product of $\overline{\overline{\sigma}}$ with a vector $\boldsymbol{A}$ is defined as being a vector, $\boldsymbol{V} = \overline{\overline{\sigma}} \cdot \boldsymbol{A}$, the a-component of which is defined by $V^a = \sigma^{am} A^m$. The total viscosity force exerted on the fluid parcel is then

$$\boldsymbol{F}_{visc} = \int_{B(V)} \overline{\overline{\sigma}} \cdot d\boldsymbol{S}_{out} \tag{31}$$

The flux-divergence theorem can be used to transform these expressions into volume integrals. This theorem holds for surface integrals of tensors as it does for vectors. To convince oneself note that, in a given reference frame, each component of these forces is given by a familiar surface integral of a vector type to which the usual form of the theorem applies. Then

$$\boldsymbol{F}_{press} + \boldsymbol{F}_{visc} = \int_V \left(-\boldsymbol{\nabla} P + \operatorname{div} \overline{\overline{\sigma}} \right) d^3 r \approx V \operatorname{div} \overline{\overline{\sigma}} - V \, \boldsymbol{\nabla} P \tag{32}$$

Let us now express the fundamental law of mechanics $\boldsymbol{F} = m\boldsymbol{\gamma}$ using expressions obtained above for the acceleration and for the forces. After simplifying the common volume factor V we are left with

$$\rho\,(\partial v/\partial t + (\boldsymbol{v} \cdot \boldsymbol{\nabla})\boldsymbol{v}) = \rho\boldsymbol{g} + \rho_e\boldsymbol{E} + \boldsymbol{j} \times \boldsymbol{B} + \boldsymbol{f}_{other} + \operatorname{div} \bar{\bar{\sigma}} - \boldsymbol{\nabla}P \qquad (33)$$

This does not look like the expected conservation equation yet. However simple manipulations reduce it to the desired form. Note, for example that

$$\rho\,\partial v/\partial t + \rho\,(\boldsymbol{v} \cdot \boldsymbol{\nabla})\boldsymbol{v} = \partial(\rho v)/\partial t + \operatorname{div}(\rho\overline{\overline{vv}}) \qquad (34)$$

This is easily shown using mass conservation, noting that, by definition, the tensor product $\overline{\overline{uv}}$ of two vectors $\boldsymbol{u}$ and $\boldsymbol{v}$ has ab-component $u^a v^b$. In particular the tensor $\rho\overline{\overline{vv}}$ has ab-component $\rho v^a v^b$. Also,

$$\boldsymbol{\nabla}P \equiv \operatorname{div}(P\,\bar{\bar{\delta}}) \qquad (35)$$

where $\bar{\bar{\delta}}$ is the unit tensor, the components δ^{ab} of which are the Kronecker symbols (zero if $a \neq b$, 1 if $a = b$). Then the fundamental law of mechanics for the fluid parcel takes the following conservative form:

$$\partial(\rho v)/\partial t + \operatorname{div}(\rho\overline{\overline{vv}} + P\,\bar{\bar{\delta}} - \bar{\bar{\sigma}}) = \rho\boldsymbol{g} + \rho_e\boldsymbol{E} + \boldsymbol{j} \times \boldsymbol{B} + \boldsymbol{f}_{other} \qquad (36)$$

The first term is the time-derivative of the momentum density, while the second is the divergence of its tensorial flux, ψ, a tensor that appears to be the sum of a convected part $\rho\mathbf{vv}$ and of microscopic momentum fluxes (the term "conductive fluxes" is not used in this case) which consist of a pressure tensor $P\delta$ and of a viscosity momentum flux tensor $-\sigma$. The sum of force densities on the r.h.s. of the equation is the volumic rate of matter momentum creation.

6.2   Density and Flux of Electromagnetic Momentum

We are familiar with the notion that the momentum of an isolated system is conserved. Isn't it surprizing then that the volumic rate of creation of momentum does not vanish, as did those of mass and charge, which are also globally conserved quantities? But indeed this is not, because the momentum for which a conservation equation has just been written is that of matter only, which is not an "isolated" system. Matter interacts with other entities such as gravity and electromagnetic fields. This really means that momentum is exchanged between the constituents of the total system, i.e matter, electromagnetic fields, photons (if they interact with matter) and gravity. It was then to be expected that the momentum of matter alone would not be globally conserved. Since electromagnetic and gravitational fields carry no mass nor charge, matter mass and electric charge are globally conserved, but this is not so for momentum. Would it then be possible to define the momentum of the electromagnetic field, of the gravity, etc.., all quantities which, added to the matter momentum would constitute a globally conserved quantity? Newtonian theory of gravitation is imperfect in this respect in that it allows propagation of gravitational signals at infinite speed. It is impossible to define for it, in that framework, a density and a flux of momentum. But the Maxwell theory of electromagnetism easily lends itself to such a definition for electromagnetic fields. It is often simpler to look at radiation from a particle point of view rather than from a field point of view, which makes it easy to define such quantities for photon populations as well. Using those Maxwell equations which involve ρ_e and j to eliminate these quantities from the expression of the Lorentz force density, then rearranging using vector calculus relations, it can be shown that:

$$\rho_e E + j \times B = -\frac{\partial}{\partial t}\left(\frac{E \times B}{\mu_0 c^2}\right) - \text{div}\left(\left(\frac{\epsilon_0 E^2}{2} + \frac{B^2}{2\mu_0}\right)\overline{\overline{\delta}} - \epsilon_0 \overline{\overline{EE}} - \frac{\overline{\overline{BB}}}{\mu_0}\right)$$
$$(37)$$

Substituting this in the matter momentum conservation equation we obtain a more global equation of conservation for material and electromagnetic forms of momentum, namely:

$$\frac{\partial}{\partial t}(\rho v + \frac{E \times B}{\mu_0 c^2})$$

$$+\text{div}\left(\rho\overline{\overline{vv}} + (P + \frac{\epsilon_0 E^2}{2} + \frac{B^2}{2\mu_0})\overline{\overline{\delta}} - \epsilon_0\overline{\overline{EE}} - \frac{\overline{\overline{BB}}}{\mu_0} - \overline{\overline{\sigma}}\right) = \rho g + f_{other} \quad (38)$$

The momentum density of the electromagnetic field is $(E \times B)/\mu_0 c^2$ and its momentum flux tensor, the Maxwell stress tensor, is

$$\overline{\overline{\psi}}_{em} = (\epsilon_0 E^2/2 + B^2/2\mu_0)\,\overline{\overline{\delta}} - \epsilon_0\overline{\overline{EE}} - \overline{\overline{BB}}/\mu_0 \quad (39)$$

The momentum conservation equation still has a gravitational source term, an unavoidable fact if indeed the system interacts with gravitation. As said above, no suitable rearrangement of this term is possible in the framework of classical theory of gravitation. If other volume forces remain on the r.h.s of the conservation equation, one may or not, depending on their nature, express them in conservative form as we did for electromagnetism. Usually this is possible because these entities often consist of particles of some type, photons, neutrinos, cosmic rays, etc.

7   Matter Internal Energy Conservation Equation

The distribution of internal energy of the fluid E_{int}, is characterized by an internal energy density $U(r, t)$, given in terms of temperature and density by an equation of state. Its flux consists as usual of a convected part, Uv, and a conducted, or microscopic, part, the heat flux q. Several effects contribute to the volumic rate of internal energy creation, such as fluid compression or expansion, viscous or Joule dissipation, radiative losses or radiative heating and so on. Expressions of the associated source terms can be precised by applying the first law of thermodynamics to a given parcel of fluid which we follow in its motion. Then:

$$dE_{int} = dQ - dT \tag{40}$$

where dQ is the amount of heat received by the parcel in dt seconds and dT is the work exerted against external world by it during the same lapse of time. This work reduces to the familiar expression PdV when viscous forces can be neglected but is otherwise more complicated. To follow the motion of the parcel, let us identify some material point inside it, M_0, which we refer to as the parcel's "center" and follow in its motion. Let v_0 be the instantaneous velocity of M_0. The velocity in the lab frame of another piece of fluid in the parcel placed at r at time t is $v_0 + c(r, t)$. Of course $c(M_0) = 0$. Note also that $\text{div} v = \text{div} c$. Since the parcel of fluid is meant to be infinitesimal, the vector field c is itself in the parcel a very small vector, scaling proportionally to the dimension ϵ of the parcel, the volume of which goes as ϵ^3. The volume V of the parcel and its boundary B vary in time. A surface element dS_{out} on B moves with respect to the center at a velocity c. It sweeps in dt seconds an (algebraic!) volume $\delta V = c dt \cdot dS_{out}$ which, by this choice of the orientation of dS is positive if the motion tends to increase the volume of the parcel and negative otherwise. The change dV of the volume V of the parcel in these dt seconds is obtained by integration on the surface B bounding V. Using the flux divergence theorem and taking advantage of the fact that V is infinitesimal, we get the volume change in the form:

$$dV = \int_{B(V)} c dt \cdot dS_{out} \approx V \, dt \, \text{div} v \tag{41}$$

The internal energy E_{int} contained in V is approximately UV . Since U and V both change in the course of time,

$$dE_{int} = V \, dU + U \, dV = V \, dU + U \, V \, dt \, \text{div} v \tag{42}$$

Let us now calculate the amount of heat received by the parcel, which consists of that part which is deposited in its volume, dQ_{int}, an algebraic quantity since both volume heating and cooling need to be considered, and of that part associated to incomes and outcomes through the border B, $dQ_{in/out}$. Note that the difference between internal and surface contributions may sometimes be rather subtle. Heating or cooling by radiation for example is the integral on surface B of the radiative energy flux, but for optically thin media, this reduces to the volume integral of the emission. Let us denote by H the volumic rate of internal heating, a positive or negative quantity. Then:

$$dQ_{int} = H \, V \, dt \tag{43}$$

The incomes and outcomes by heat conduction are given in terms of the heat flux q by:

$$dQ_{in/out} = \int_{B(V)} q \cdot dS_{in} \approx -V \, dt \, \mathrm{div} q \qquad (44)$$

Let us now calculate the work exerted by the parcel on the external world in dt seconds. It consists of the work of volume and contact forces. A moment's thought shows that the work of the former, calculated in the rest frame which instantaneously accompanies the parcel's motion, is negligible because it is of order of the product $|c||f|V$ where c is the velocity with respect to the parcel's center, of order ϵ, f is the density of force and V the volume, of order ϵ^3. The result is $\mathcal{O}(\epsilon^4)$, one order in ϵ larger than other retained terms. For example the work of surface forces is typically of order $dt|c|PS$ where S is the surface of B, which scales as ϵ^2, so that the surface force term scales as ϵ^3. To be specific, the work exerted by the forces that the fluid of the parcel exerts at its boundary on the outer world is

$$dT = \int_{B(V)} c dt \cdot \left(P \, dS_{out} - \overline{\overline{\sigma}} \cdot dS_{out}\right) \qquad (45)$$

The dot-product $c \cdot \overline{\overline{\sigma}}$ is a vector, the i-component of which is $c^j \sigma^{ji}$. Using the flux-divergence theorem and developing $\mathrm{div}(Pc)$, we obtain, for an infinitesimal volume V:

$$dT = V dt \left(\mathrm{div}(Pc) - \nabla_i(c^j \sigma^{ji})\right) =$$

$$V dt \left(P\mathrm{div}c + c \cdot \nabla P - (\nabla_i c^j)\sigma^{ji} - c^j(\nabla_i \sigma^{ji})\right) \qquad (46)$$

Terms having factors proportional to c are negligible, being $\mathcal{O}(\epsilon^4)$. The partial derivatives of $c(r,t)$ with respect to space are not small, though, and equal similar derivatives of v. In particular div $c = $ div v. Thus

$$dT = V \, dt \left(P\mathrm{div} \, v - \sigma^{ji}(\nabla_i v^j)\right) \qquad (47)$$

which, in geometrical terms, expresses as:

$$dT = V \, dt \left(P\mathrm{div} \, v - (\overline{\overline{\sigma}} \cdot \nabla) \cdot v\right) \qquad (48)$$

So, the first law of thermodynamics applied to the fluid parcel takes the form:

$$V \, dU + V dt \, U \mathrm{div} v = (HV dt - V dt \, \mathrm{div} q) - V dt \left(P\mathrm{div} v - (\overline{\overline{\sigma}} \cdot \nabla) \cdot v\right) \quad (49)$$

Using the further relation $dU = (dU/dt)dt$ where (dU/dt) is meant to be the lagrangean derivative of U, and simplifying by $V dt$, we obtain:

$$\partial U/\partial t + \mathrm{div} \left(U \, v + q\right) = H - P\mathrm{div} v + (\overline{\overline{\sigma}} \cdot \nabla) \cdot v \qquad (50)$$

Once more, this equation is not perfectly conservative. This is no surprise, since it expresses fluid internal energy balance only. But this form of energy is exchanged in the interaction process with other forms of energy, such as kinetic energy of organized motions, electromagnetic energy, etc... The volumic rate of creation which is seen on the r.h.s of equation (50) describes these exchanges. As for momentum, it is possible to write down conservation equations for more global forms of energy, for example fluid internal and kinetic energy, or internal, kinetic, and electromagnetic energy, etc.. The next chapter is devoted to establishing these conservation equations.

8   Conservation Equations for More Forms of Energy

8.1   Fluid Internal and Kinetic Energy Conservation Equation

To form the conservation equation for kinetic energy of organized (as opposed to thermal) motions, the density of which is $\rho v^2/2$, we dot-multiply the equation of motion by v. Some further algebraic manipulations reduce this, using mass conservation, to:

$$\frac{\partial}{\partial t}\left(\frac{1}{2}\rho v^2\right) + \text{div}\left(\frac{1}{2}\rho v^2\,v\right) =$$

$$-v\cdot\nabla P + v\cdot\text{div}\overline{\overline{\sigma}} + \rho v\cdot g + v\cdot(\rho_e E + j\times B) + v\cdot f_{other} \qquad (51)$$

Adding this with the conservation equation (50) for the internal energy of the fluid we get, after simple algebra:

$$\frac{\partial}{\partial t}\left(\frac{1}{2}\rho v^2 + U\right) + \text{div}\left(\frac{1}{2}\rho v^2\,v + (U+P)v + q - v\cdot\overline{\overline{\sigma}}\right) =$$

$$\rho v\cdot g + v\cdot f_{other} + v\cdot(\rho_e E + j\times B) + H \qquad (52)$$

In the first term we recognize the kinetic and internal energy density of the fluid, and under operator div, in the second term, we see the flux of internal and kinetic energy which consists of a convected kinetic energy flux, an enthalpy flux (enthalpy is $E_{ent} = E_{int} + PV$ and its density is $h = U + P$), the heat flux , q, and the energy flux associated to viscosity, $-v\cdot\overline{\overline{\sigma}}$. The source terms on the right hand side represent exchanges of energy with other forms of it, electromagnetic in particular.

8.2   Internal, Kinetic and Electromagnetic Energy Conservation Equation

It is again possible to transform the source terms associated to exchanges between material and electromagnetic energy so that they appear in conservative form, and constitute with the other terms a conservation equation for the sum of material and electromagnetic energy. To achieve this, the Joule heating term, which participates exchanges between these forms of energy

must be singled out of the general volume heating term H. We then write it
as

$$H = H_{joule} + H_{other} = j_{fluid}^2/\sigma_e + H_{other} \tag{53}$$

where j_{fluid} is the electric current density as observed in the rest frame where
the piece of fluid located at r is instantly at rest. It is given by Ohm's law,
$j_{fluid} = \sigma_e \, E_{fluid}$. It has been asumed here that the electrical conductivity
σ_e is isotropic. The vector E_{fluid} is the electric field observed in the fluid's
rest frame. In the galilean approximation, it is related to the electric and
magnetic fields in the lab frame, E and B, by

$$E_{fluid} = E + v \times B \tag{54}$$

The electric current density in the lab frame is, on the other hand, given by:

$$j = j_{fluid} + \rho_e v \tag{55}$$

from which it is deduced that

$$H_{joule} = j_{fluide} \cdot (E + v \times B) \tag{56}$$

$$H_{joule} + v \cdot (\rho_e E + j \times B) = j_{fluide} \cdot (E + v \times B) + \rho_e E \cdot v + v \cdot ((j_{fluide} + \rho_e v) \times B) \tag{57}$$

After some simplifications owing to vanishing or identical mixed products we
are left with:

$$H_{joule} + v \cdot (\rho_e E + j \times B) = (\rho_e v + j_{fluide}) \cdot E = j \cdot E \tag{58}$$

Using Maxwell's equations $j \cdot E$ can finally be reduced to a conservative form:

$$j \cdot E = -\frac{\partial}{\partial t} \left(\frac{\epsilon_0 E^2}{2} + \frac{B^2}{2\mu_0} \right) - \text{div}(\frac{E \times B}{\mu_0}) \tag{59}$$

Inserting this in the material energy conservation equation, we get:

$$\frac{\partial}{\partial t} \left(\frac{1}{2}\rho v^2 + U + \frac{1}{2}\epsilon_0 E^2 + \frac{1}{2}\frac{B^2}{\mu_0} \right) +$$

$$\text{div} \left(\frac{1}{2}\rho v^2 v + (U + P)v + q - v \cdot \overline{\overline{\sigma}} + \frac{E \times B}{\mu_0} \right)$$

$$= \rho v \cdot g + v \cdot f_{other} + H_{\neq joule} \tag{60}$$

8.3   Internal, Kinetic, Gravitational and Electromagnetic Energy Conservation Equation

Further progress can be achieved in globalizing the forms of energy included
in the balance if the gravity in which the plasma moves is independent of time.

This is consistent with actual fluid motion if self-gravitation is negligible. Let G be the gravitational potential. The following relations hold true, the last one implying time independence of G:

$$\rho v \cdot g = -\rho v \cdot \nabla G = -(\text{div}(\rho v G) - G \text{div} \rho v)$$

$$= -\text{div}(\rho v G) - G(\frac{\partial \rho}{\partial t}) = -\frac{\partial}{\partial t}(\rho G) - \text{div}(\rho v G) \tag{61}$$

A conservation equation for the internal, kinetic, electromagnetic and gravitational forms of energy is thus obtained in the form:

$$\frac{\partial}{\partial t}\left(\frac{1}{2}\rho v^2 + U + \frac{1}{2}\epsilon_0 E^2 + \frac{1}{2}\frac{B^2}{\mu_0} + \rho G\right) +$$

$$\text{div}\left(\frac{1}{2}\rho v^2 v + (U + P)v + q - v \cdot \overline{\overline{\sigma}} + \frac{E \times B}{\mu_0} + \rho G v\right) = H_{\neq joule} + v \cdot f_{\neq g, E, B} \tag{62}$$

9   A Provisional Synthesis

To sum up, the equations of MHD consist of

- The three conservation equations for mass, momentum and energy.
- The Maxwell equations which determine the electric variables.
- The necessary equations of state, like $P(\rho, T)$ and $U(\rho, T)$.
- Phenomenological relations which give the microscopic fluxes.

These are the heat flux q, the conduction electric current density, j_{cond} and the viscous stress tensor $\overline{\overline{\sigma}}$. Physical laws, valid when the fluid is everywhere near a state of thermodynamic equilibrium, are deduced from kinetic theory of gases and plasmas (see part II) to give them in terms of gradients of MHD variables, such as density, fluid velocity, temperature, magnetic and electric field. One such law is, for example, Fourier's heat conduction law which gives the heat flux in terms of the temperature gradient, another one is Ohm's law which gives the conducted electric current density in terms of the electric field.

10   Subrelativistic Limit

Considerable simplification is obtained in the limit of sub-relativistic motions, an assumption which was already implicit in our use of the fundamental law of newtonian mechanics. We then now consider the fluid velocity to be much less than the speed of light, c, and the phase velocity of waves which propagate in the fluid to be also small as compared to it. We recall that the square of the speed of light is $c^2 = 1/(\epsilon_0 \mu_0)$ where ϵ_0 is the dielectric permittivity of vacuum and μ_0 its magnetic permeability. Let T be the characteristic time over which the MHD variables vary and L be the characteristic scale length of these variations. In sub-relativistic conditions as defined above the ratio $V \equiv L/T$ is of order of the fluid velocity or of order of the phase velocity of MHD waves propagating in this plasma. Assume both to be much less than the speed of light, c. The temporal derivative of any quantity X being estimated, in rough order of magnitude, to be comparable to X/T, and any of its partial derivative with respect to space coordinates, whichever operator it is involved in, being roughly estimated as comparable to X/L, it is easily found, from Faraday equation $\mathrm{rot}\,\boldsymbol{E} = -\partial\boldsymbol{B}/\partial t$ that the ratio $E/B \approx V$. This estimate is incorrect only in the exceptional situation when the fluid is placed in a very intense external electrostatic (i.e curl free) field. Assume that this is not so. From the above result we can deduce that the displacement current $\epsilon_0 \partial\boldsymbol{E}/\partial t$ is negligible compared to $\mathrm{rot}\,\boldsymbol{B}/\mu_0$. Indeed their ratio is of order $(\epsilon_0 E/T)/(B/\mu_0 L)$ which is about V^2/c^2. The Maxwell's equations can then be simplified by reduction to the magnetostatic approximation, in which the displacement current is ignored. The convected electric current $\rho_e \boldsymbol{v}$ then turns out to be negligible compared to the total electric current, $\boldsymbol{j}$, because their ratio is $(\epsilon_0 \,\mathrm{div}\,E)V/(\mathrm{rot}\,\boldsymbol{B}/\mu_0)$, which is again of order V^2/c^2. This implies that the electric current essentially reduces to the conduction current. A similar estimate shows that the density of electric force, $\rho_e \boldsymbol{E}$ is negligible to the Lorentz force density, $\boldsymbol{j} \times \boldsymbol{B}$. Finally note that the electric energy density is negligible to the magnetic energy density, their ratio being $(\epsilon_0 E^2)/(B^2/\mu_0) \approx V^2/c^2$. As a result the electric charge density has disappeared from all but the Poisson's equation, $\epsilon_0 \,\mathrm{div}\,\boldsymbol{E} = \rho_e$. The latter is then of no use in MHD, unless one is tempted, just for the sake of it, to calculate the charge density.

11 Synthesis of Subrelativistic MHD Equations

The variables of subrelativistic MHD are then the mass density ρ, the fluid velocity, $\boldsymbol{v}$, the temperature T and the electric and magnetic fields, $\boldsymbol{E}$ and $\boldsymbol{B}$. These variables obey the system of the three conservation equations of MHD coupled to Maxwell's equations (save the unuseful Poisson equation). Auxiliary quantities appear in these equations, like pressure and internal energy density, given as a functions of ρ and T by equations of state, and microscopic fluxes. For example the heat flux $\boldsymbol{q}$, the conduction current, $\boldsymbol{j}$ or the microscopic momentum flux, i.e. the viscous stress tensor $\overline{\overline{\sigma}}$. These fluxes are given by transport laws as a function of MHD variables and their gradients. The transport coefficients which enter these relations are the coefficient of thermal conductibility χ, the electric conductibility σ_e, and the dynamic viscosity coefficient η. The kinetic theory of plasmas near thermodynamic equilibrium allowed to calculate them, once and for all, as functions of ρ and T. The gravitational potential G, the gravity field $\boldsymbol{g}$, the forces other than those explicitly written in the momentum equation and the volumic heating rates others than those explicitly written in the energy equation, if present, must be calculated by independent complementary theories. For example newtonian theory of gravity if the system is self-gravitating or radiative transfer theory if momentum and energy are exchanged with the photon field. The three conservation equations of MHD are:

$$\frac{\partial \rho}{\partial t} + \text{div}(\rho \boldsymbol{v}) = 0 \tag{63}$$

$$\rho \left(\frac{\partial \boldsymbol{v}}{\partial t} + (\boldsymbol{v} \cdot \boldsymbol{\nabla}) \boldsymbol{v} \right) = \rho \boldsymbol{g} - \boldsymbol{\nabla} P + \text{div}\overline{\overline{\sigma}} + \boldsymbol{j} \times \boldsymbol{B} + \boldsymbol{f}_{other} \tag{64}$$

$$\frac{\partial}{\partial t} \left(U + \frac{1}{2}\rho v^2 + \rho G + \frac{B^2}{2\mu_0} \right)$$
$$+\text{div} \left(\frac{1}{2}\rho v^2 \boldsymbol{v} + (U + P)\boldsymbol{v} + \boldsymbol{q} - \boldsymbol{v} \cdot \overline{\overline{\sigma}} + \rho G \boldsymbol{v} + \frac{\boldsymbol{E} \times \boldsymbol{B}}{\mu_0} \right) = H_{other} + \boldsymbol{v} \cdot \boldsymbol{f}_{other} \tag{65}$$

The three useful Maxwell equations, simplified for sub-relativistic situation, are:

$$\text{div}\boldsymbol{B} = 0 \tag{66}$$

$$\text{rot}\boldsymbol{E} = -\partial \boldsymbol{B}/\partial t \tag{67}$$

$$\text{rot}\boldsymbol{B} = \mu_0 \boldsymbol{j} \tag{68}$$

Equations of state give pressure and internal energy density:

$$P = P(\rho, T) \qquad\qquad U = U(\rho, T) \tag{69}$$

Transport laws give the microscopic fluxes of heat, charge and momentum as:

$$\boldsymbol{q} = -\chi(\rho, T)\boldsymbol{\nabla}T \tag{70}$$

$$\boldsymbol{j} = \sigma_e(\rho, T)(\boldsymbol{E} + \boldsymbol{v} \times \boldsymbol{B}) \tag{71}$$

$$\sigma_{ij} = \eta(\rho, T) \left(\nabla_i v_j + \nabla_j v_i - \frac{2}{3}\delta_{ij}\text{div}\boldsymbol{v} \right) \tag{72}$$

12   Magnetic Pressure and Tension

The Lorentz force is sometimes presented as being the sum of the gradient of a magnetic pressure and of a magnetic tension force, which is justified as follows. The Lorentz force density can be expressed in terms of the magnetic field alone as:

$$\boldsymbol{j} \times \boldsymbol{B} = \frac{1}{\mu_0}\text{rot}\boldsymbol{B} \times \boldsymbol{B} \tag{73}$$

By vector calculus identities, this is transformed into:

$$\frac{1}{\mu_0}\mathrm{rot}B \times B = -\nabla \frac{B^2}{2\mu_0} + \frac{1}{\mu_0}\left(B \cdot \nabla\right)B \tag{74}$$

The first of these two terms is the gradient of some pressure, the magnetic pressure, while the second is the so-called magnetic tension. So written, however, the basic fact that the total Lorentz force is perpendicular to the magnetic field is no more apparent. We can work out a somewhat more sophisticated expression by writing the magnetic field as $B = Bt$ where B is its modulus and t is a unit vector tangent to the field line. Then,

$$(B \cdot \nabla)B = Bt \cdot \nabla(Bt) = t\left((t \cdot \nabla)(\frac{B^2}{2})\right) + B^2\left((t \cdot \nabla)t\right) \tag{75}$$

Using Frenet formulae for curvature and torsion of curves,

$$(t \cdot \nabla)t \equiv \frac{dt}{ds} = \frac{N}{R_c} \tag{76}$$

where N is the unit vector in the direction of the principal normal to the field line and R_c is its radius of curvature. Gathering these results,

$$j \times B = -\left(\nabla \frac{B^2}{2\mu_0} - t(t \cdot \nabla)\frac{B^2}{2\mu_0}\right) + \left(\frac{B^2}{\mu_0}\right)\frac{N}{R_c} =$$

$$-\left(\overline{\overline{\delta}} - \overline{\overline{tt}}\right) \cdot \nabla \frac{B^2}{2\mu_0} + \left(\frac{B^2}{\mu_0}\right)\frac{N}{R_c} \tag{77}$$

The first term on the right hand side is again the gradient of the magnetic pressure, from which the projection along the field line has however been substracted. This substraction reduces it to its part perpendicular to the field line. Let us name it the "perpendicular" gradient of magnetic pressure. The second term, also perpendicular to B, grows larger when the radius of curvature of the field line becomes smaller. It is this term which is properly the "magnetic tension" force. This separation of the Lorentz force into two parts should not obscure the fact that they are always associated. It would be incorrect for example to estimate the magnitude of magnetic forces on the basis of magnetic pressure alone: if the magnetic field is potential (zero electric current density, or equivalently $\mathrm{rot}B = 0$) the net force is zero, the gradient of magnetic pressure being in this case exactly compensated by magnetic tension.

13   MHD Self-consistency

It is interesting to observe how MHD variables are coupled. An MHD system is a motor which produces its own electricity and to some extent its own magnetic field. Indeed the magnetic field and the electric current (which, to within an insignificant factor, is just the rotational of the former) interact (couple) to produce the Lorentz force $j \times B$. This, together with other forces like gas pressure gradients, determine the fluid motion, v. This is the motor effect. Now, the fluid motion couple to the magnetic field to produce the electromotive field $v \times B$, which, by Ohm's law, contributes to the generation of an electric current circulating in the plasma and hence to the associated magnetic field. This is the "electro-motive", or dynamo, effect. This coupling between the hydrodynamic quantities and the magnetic field may be tight and complex. The situation simplifies when the Lorentz force has but a negligible influence on the plasma motion, which happens when $|j \times B| \ll |\nabla P|$. Assuming that the gradient scale of pressure and magnetic field are comparable, this reduces to the condition that the plasma "β parameter", the ratio of gas to magnetic pressure, be large. An opposite limit is when the magnetic forces potentially dominate all other forces. They dominate over gas pressure forces if $\beta \ll 1$, an inequality which can also be written as $(P/\rho) \ll (B^2/(\mu_0\rho))$, comparing the square of the sound speed $c_s^2 = \gamma P/\rho$ to $B^2/(\mu_0\rho)$. Magnetic forces dominate over inertia forces when $|\rho(v \cdot \nabla)v| \ll |j \times B|$ or equivalently when v^2 is much less than $B^2/(\mu_0\rho)$. The latter quantity is the square of a characteristic velocity which appears ubiquitously in MHD, the Alfven velocity c_A:

$$c_A^2 = \frac{B^2}{\mu_0\rho} \tag{78}$$

So, when $|v| \ll c_A$, inertia forces are negligible to the Lorentz forces, and when $c_s \ll c_A$, the gas pressure forces are negligible to the Lorentz forces. When both inequalities are satified the dominant term in the equation of motion becomes the $j \times B$ force alone. This equation cannot be satisfied, unless this vector product vanishes. This conclusion is obtained when neglecting all little pressure and inertia terms, i.e. it is the solution correct to zeroth order in (c_s/c_A) and (v/c_A). A magnetic structure such that everywhere in some region $j \times B = 0$ is said to be "force free" in this region, this meaning "free of Lorentz forces".

14   Equation of Evolution of the Magnetic Field

14.1   Field Evolution

Using Faraday's and Ampere's equations in conjunction with Ohm's law we obtain the time variation of the magnetic field:

$$\frac{\partial B}{\partial t} = \text{rot}(v \times B) - \text{rot}\left(\frac{1}{\mu_0\sigma_e}\text{rot}B\right) \tag{79}$$

Often the magnetic diffusivity $\eta_m = (\mu_0 \sigma_e)^{-1}$ is treated as constant in space. In this case the field evolution equation reduces to:

$$\frac{\partial \boldsymbol{B}}{\partial t} = \text{rot}(\boldsymbol{v} \times \boldsymbol{B}) + \eta_m \Delta \boldsymbol{B} \qquad (80)$$

The justification for accepting this simplification is that often, as we shall see, the second term is either negligible or significant in restricted regions of space only. So it does not matter very much if the variations in space of η_m is carefully taken care of or not, the phenomena associated with this dissipative effect remaining qualitatively the same.

14.2   The Kinematic and Dynamic Dynamo Problems

At first sight the field evolution equation (80) looks linear in $\boldsymbol{B}$. However, it should be borne in mind that the velocity field $\boldsymbol{v}$ depends implicitly on $\boldsymbol{B}$ because the Lorentz force takes part in determining the motion. The linearity is then only apparent but not real, unless it happens that the motion is in fact only weakly dependent on these forces, i.e. $\beta \gg 1$. In this case the dynamo problem, which consists in finding how, starting from some initial very small seed, the magnetic field is amplified to finite amplitude by motions of the conducting fluid, reduces to linearity. A plasma flow, often a stationary one, would then be calculated or assumed and the associated solutions of the field evolution equation (80) would be analyzed in eigenmodes of its right hand side. There is dynamo action if one of these modes is associated with an eigenvalue that causes temporal growth. This is the "kinematic", or linear, dynamo problem. No reaction of the Lorentz forces against the motion is taken into account at this stage. Because dynamo action may be associated with turbulent flows, an interesting variant of the kinematic dynamo problem consists in treating the velocity field as a random vector function of known statistical properties and try and calculate whether there is field growth, on which scales and which are the statistical properties of this field. However, since the solutions which show dynamo action have exponential field growth, the reaction of Lorentz forces must eventually be taken into account to find the saturation state, if any, or the further development in time of the motion, which may be oscillating or chaotic. This defines the dynamic, or non-linear, dynamo problem. Obviously, only the electromotive term, $\text{rot}(\boldsymbol{v} \times \boldsymbol{B})$ can cause dynamo action because the other term, which pictures Joule dissipation, gives rise to a linear diffusion equation unable of any instability.

One may wonder why Joule dissipation gives rise to diffusive effects, weakening field gradients as described by the diffusion equation, rather than simply weakening the electric currents initially present in the system. This in fact is due to the combined effect of electrical resistivity and Lenz's law. When, because of resistivity, the electric current weakens somewhere, the magnetic field which it generates also weakens, which causes an induced electric field to appear (Faraday law). This field drives Lenz's currents even at places where there might have previously been none. So, electric currents weaken where strong but increase where weak. The result of these combined resistive and inductive effects is a diffusion of the electric currents: this can be clearly seen by exerting the rot operator on the field evolution equation in the absence of electromotive term, which yields a diffusion equation for the electric current density.

14.3 The Magnetic Reynolds Number and the Perfect MHD Limit

When both the electromotive and dissipative terms are present in the field evolution equation, which one dominates? Let L be a characteristic scale of the magnetic field gradients. The ratio of these terms is approximately

$$|\text{rot}(\boldsymbol{v} \times \boldsymbol{B})|/|\eta_m \Delta \boldsymbol{B}| \approx (vB/L)/(\eta_m B/L^2) = vL/\eta_m \tag{81}$$

The dimensionless quantity

$$R_m = vL/\eta_m \tag{82}$$

is the magnetic Reynolds number. Because of the large sizes of astrophysical objects and of the good electric conductibility of gaseous plasmas, the magnetic Reynolds number of MHD flows in such objects is usually very large. It is of order 10^{14} in the solar corona, for a temperature of 10^6 K, $L = 1R_\odot$ and $v = 100$ km/s. One can then think of considering the limit of infinite R_m. We discuss below what could be lost so doing. The limit in which plasma resistivity and viscosity (also a dissipative effect) are neglected is called perfect MHD. In this limit the field evolves under the electromotive term alone, as described by the "perfect MHD induction equation":

$$\frac{\partial \boldsymbol{B}}{\partial t} = \text{rot}(\boldsymbol{v} \times \boldsymbol{B}) \tag{83}$$

In this limit the electrical resistivity σ_e^{-1} is regarded as strictly zero, as is also j/σ_e, and Ohm's law simplifies to:

$$\boldsymbol{E} + \boldsymbol{v} \times \boldsymbol{B} = 0 \tag{84}$$

from which equation (83) follows by taking its rotational. Equation (84) physically means that the electric field vanishes in the fluid instantaneous frame of rest: the very good plasma conductor finds at any time a state of electrostatic equilibrium in its own rest frame.

15 The Flux and Field Freezing Theorems

In the perfect MHD limit the plasma behaves like a super-conductor, not in the quantum-physical sense of this term, but in its electric sense of absolute zero resistivity. The field evolution then has a number of interesting properties.

(a) The flux through any circuit moving with the fluid remains constant in time.

(b) Fluid elements connected by a field line at some time remain so later on. These results are known as "flux freezing" or "field freezing" theorems, a term which refers to the fact that matter and field accompany eachother in the motion. Field lines appear as "frozen- in" in the plasma (or conversely, the plasma appears as "frozen-in" onto field lines) Let us prove the first of these two theorems.

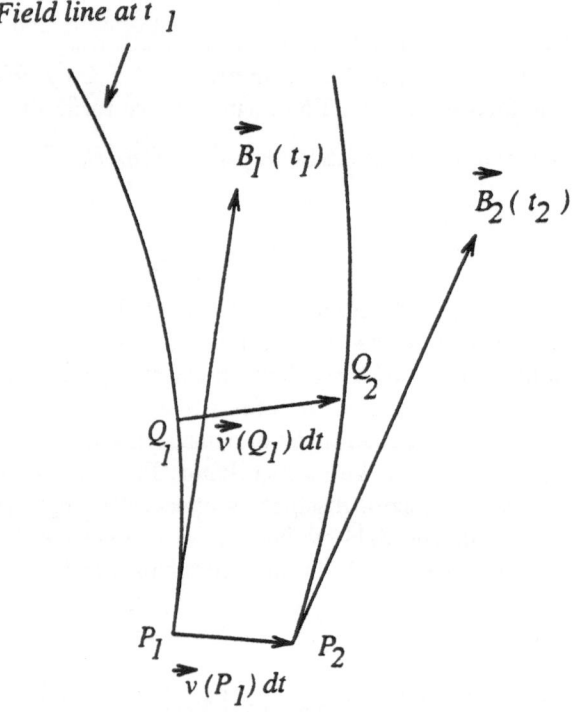

Fig. 1. A circuit C_1 moving with the fluid becomes the circuit C_2 a time dt later. The surface made of the union of the two surfaces enclosed by C_1 and C_2 and the lateral surface Σ made of fluid displacements from one to the other is a closed surface, S. Normal elements oriented outwards to S are shown.

15.1   Proof of Perfect MHD Flux Conservation
Following the Motion

Consider at time t_1 a circuit C_1, each point of which, M_1 say, accompanies the local fluid in its motion. At time $t_2 = t_1 + dt$, the point M_1 has reached a position M_2 given by

$$OM_2 = OM_1 + v(M_1, t_1)\, dt = OM_1 + dM_1 \qquad (85)$$

The locus of all points M_2 defines the circuit C_2, displaced from C_1 by the plasma motion. Note that in this expresssion we need not, to first order in dt, distinguish between $\boldsymbol{v}(M_1, t_1)dt$ and $\boldsymbol{v}(M_2, t_2)dt$ henceforth noted as $\boldsymbol{v}(M)dt$. Let S be the closed surface consisting of a surface $S(C_1)$ spanning C_1, another surface $S(C_2)$ spanning C_2 and the surface Σ generated by the set of all vectors $\boldsymbol{v}(M)dt$ for all M's on C_1, joining C_1 to C_2 (Fig. 1).

The magnetic flux at time t_2 through S is zero:

$$\int_S \boldsymbol{B}(P, t_2) \cdot d\boldsymbol{S}_{out} = 0 \qquad (86)$$

In this integral the surface element $d\boldsymbol{S}_{out}$ is oriented outwards with respect to the volume enclosed by S. This surface integral consists of three parts:

$$\int_S \boldsymbol{B}(P, t_2) \cdot d\boldsymbol{S}_{out} = \int_{S(C_1)} \boldsymbol{B}(P, t_2) \cdot d\boldsymbol{S}_{out}$$

$$+ \int_{S(C_2)} \boldsymbol{B}(P, t_2) \cdot d\boldsymbol{S}_{out} + \int_\Sigma \boldsymbol{B}(P, t_2) \cdot d\boldsymbol{S}_{out} \qquad (87)$$

The second term on the right hand side is Φ_2, the flux at time t_2 through C_2. The first term is not the flux Φ_1 at time t_1 through C_1 for two reasons. One is that the value of $\boldsymbol{B}$ involved in it is taken at time t_2. The other is that the surface element $d\boldsymbol{S}_{out}$ is oriented on the surface $S(C_1)$ opposite (with respect to a common circulation sense on C_1 and C_2) to what it is on $S(C_2)$. Correct to first order in dt we can write:

$$\boldsymbol{B}(t_2) = \boldsymbol{B}(t_1) + \frac{\partial \boldsymbol{B}}{\partial t} dt = \boldsymbol{B}(t_1) + \text{rot}(\boldsymbol{v} \times \boldsymbol{B})dt \qquad (88)$$

obtaining from (86) and (87)

$$0 = \Phi_2 + \int_{S(C_1)} \boldsymbol{B}(P, t_1) \cdot d\boldsymbol{S}_{out} + dt \int_{S(C_1)} \text{rot}(\boldsymbol{v} \times \boldsymbol{B}) \cdot d\boldsymbol{S}_{out} + \int_\Sigma \boldsymbol{B}(P, t_2) \cdot d\boldsymbol{S}_{out}$$
$$(89)$$

Because of the different orientation of the normal on $S(C_2)$ and $S(C_1)$ the second term on the right hand side is $-\Phi_1$. The other two are of order dt, the last one because the surface Σ between C_1 and C_2 is generated by infinitesimal vectors $\boldsymbol{v}\, dt$. Let $d\boldsymbol{l}$ be the line element in the vicinity of point M on C_1, oriented in the direct sense to the direction of the normal vector defining the flux Φ_1. The outgoing surface element on Σ can be written as

$$d\boldsymbol{S}_{out} = d\boldsymbol{l} \times \boldsymbol{v}(M)dt \qquad (90)$$

Neglecting terms quadratic in dt we then obtain

$$\Phi_1 - \Phi_2 = dt \int_{S(C_1)} \text{rot}(\boldsymbol{v} \times \boldsymbol{B}) \cdot d\boldsymbol{S}_{out} + dt \int_{C_1} \boldsymbol{B}(M) \cdot (d\boldsymbol{l} \times \boldsymbol{v}(M)) \quad (91)$$

By Stokes theorem, the surface integral can be transformed into a line integral on C_1. Paying due attention to the fact that the outwards normal on $S(C_1)$

is opposite to the sense with respect to which flux Φ_1 is defined, we obtain:

$$\Phi_1 - \Phi_2 = -dt \int_{C_1} (v(M) \times B(M)) \cdot dl + dt \int_{C_1} B(M) \cdot (dl \times v(M)) \quad (92)$$

The sum on the r.h.s. is zero, because the mixed products $(v \times B) \cdot dl$ and $B \cdot (dl \times v)$ are equal. This then gives, as previously claimed, $\Phi_2 = \Phi_1$.

15.2 Proof of the Permanence of Magnetic Connection in Perfect MHD Motions

Let us now prove the second theorem. Consider two fluid elememts which are, at time t_1, at points P_1 and Q_1, at an infinitesimal distance from eachother on a common field line. The vector P_1Q_1 is then parallel to $B(P_1, t_1)$, which is expressed by

$$P_1Q_1 \times B(P_1, t_1) = 0 \quad (93)$$

At time $t_2 = t_1 + dt$, the first plasma element has moved to a point P_2 and the second to a point Q_2. The theorem claims that $P_2Q_2 \times B(P_2, t_2)$ is also zero (Fig. 2).

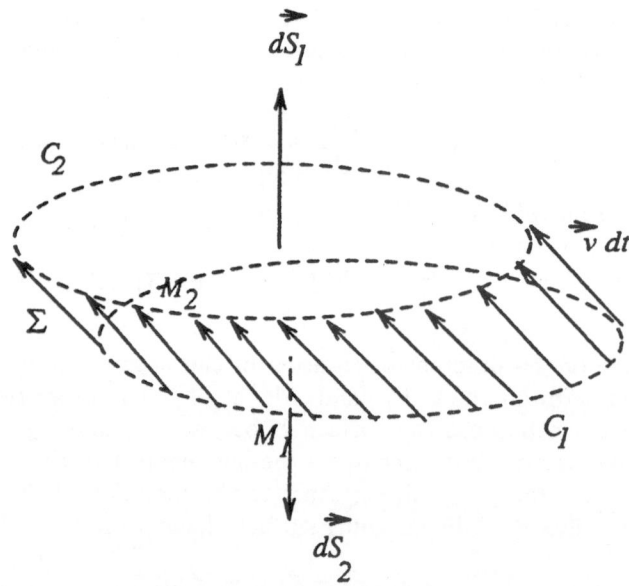

Fig. 2. Two neighbouring fluid elements P_1 and Q_1 on a common field line at time t_1 are moved, dt seconds later, at P_2 and Q_2. Under perfect MHD conditions, they still are on a common field line at that time. See text for a proof.

Correct to first order in dt,

$$OP_2 = OP_1 + v(P_1, t_1)dt \quad (94)$$

$$OQ_2 = OQ_1 + v(Q_1, t_1)dt \tag{95}$$

The field at P_2 at time t_2 is given in terms of $B(P_1, t_1)$ by

$$B(P_2, t_2) = (B(P_2, t_2) - B(P_2, t_1)) + (B(P_2, t_1) - B(P_1, t_1)) + B(P_1, t_1) \tag{96}$$

Expanding to first order in dt and $dP \equiv P_1 P_2 = v(P_1) \, dt$:

$$B(P_2, t_2) = B(P_1, t_1) + \frac{\partial B}{\partial t} \, dt + (dP \cdot \nabla)B \tag{97}$$

A first order Taylor expansion of each component of B with respect to the coordinates of point P gives:

$$B(P + dP) = B(P) + (dP \cdot \nabla)B \tag{98}$$

The vector $P_2 Q_2$ can be similarly expanded:

$$P_2 Q_2 = P_1 Q_1 + (v(Q_1) - v(P_1)) \, dt = P_1 Q_1 + (P_1 Q_1 \cdot \nabla)v(P) \tag{99}$$

Let us note for conciseness

$$K_1 = P_1 Q_1 \qquad K_2 = P_2 Q_2 \qquad B_1 = B(P_1, t_1) \qquad B_2 = B(P_2, t_2) \tag{100}$$

we get:

$$K_2 \times B_2 = (K_1 + (K_1 \cdot \nabla)v(P) \, dt) \times (B_1 + dt(v \cdot \nabla)B + dt \, \mathrm{rot}(v \times B)) \tag{101}$$

and, correct to first order in dt:

$$K_2 \times B_2 - K_1 \times B_1 = K_1 \times (dt(v \cdot \nabla)B + dt \, \mathrm{rot}(v \times B)) + (K_1 \cdot \nabla)v(P)dt \times B_1$$

A vector calculus identity states that:

$$\mathrm{rot}(v \times B) = (\mathrm{div}B) \, v - (\mathrm{div}v) \, B + (B \cdot \nabla)v - (v \cdot \nabla)B \tag{103}$$

whence:

$$K_2 \times B_2 - K_1 \times B_1 = K_1 \, dt \times ((v \cdot \nabla)B - (\mathrm{div}\ v)B + (B \cdot \nabla)v - (v \cdot \nabla)B)$$

$$+ dt \, ((K_1 \cdot \nabla)v) \times B \tag{104}$$

Simplifying:

$$K_2 \times B_2 - K_1 \times B_1 = dt \, (K_1 \times ((B \cdot \nabla)v - (\mathrm{div}v) \, B) + ((K_1 \cdot \nabla)v) \times B) \tag{105}$$

It is assumed that K_1 and B_1 are parallel. Let t be the unit vector along their common direction and B_1 and K_1 be their moduli:

$$B_1 = B_1 t \qquad\qquad K_1 = K_1 t \tag{106}$$

The difference $K_2 \times B_2 - K_1 \times B_1$ reduces to:

$$K_2 \times B_2 - K_1 \times B_1 = K_1 B_1 \, dt \, (t \times ((t \cdot \nabla)v - (\mathrm{div}v) \, t) + ((t \cdot \nabla)v) \times t) \tag{107}$$

The right hand side of equation (107) is zero, which is just what the theorem claims.

15.3   Consequences and Limits of the Flux Freezing Theorems

In perfect MHD flows, field lines accompany the fluid. One could say that they are therefore materialized and are so to speak unbreakable because they always link the same fluid elements. As a result their topology cannot change. For example, fluids elements which are not initially on a common field line cannot become linked by one later on, at least as long as the perfect MHD properties apply. This general topological constraint restricts the perfect-MHD motions, forbidding a lot of movements that would otherwise appear to be perfectly conceivable. Conversely, the constraint that the field follows the fluid motion, whatever its complexity, may create situations where the magnetic structure becomes itself very complex, the rotational of the magnetic field, proportional to the electric current density, becoming non-zero and of a very complex structure too. Consider for example the sun (Fig. 3), which, as is known, does not rotate near its surface like a solid body, the equatorial regions rotating faster. Suppose that initially the field lines run just below the surface in meridian planes, and emerge at some northern and southern latitudes. From the second magnetic line freezing theorem, they are transported by the matter.

Assume, just for simplicity that the reaction of Lorentz forces against the rotation motion can be neglected. Consider plasma elements which at the initial time are on such a common, poloidal, field line. Because the rotation of polar elements lags over that of equatorial ones, this field line does not remain in a plane, but bulges azimuthally more and more as the motion goes on. The field line tends to wrap around the sun, developing a growing azimuthal field component of one sign in one hemisphere and of another sign in the other. This component results from a system of poloidal electric currents which are generated by the electromotive field of the differential rotation motion. When field lines become very tightly wrapped azimuthally it becomes no longer possible to neglect the reaction on the fluid motion of the growing associated Lorentz force. Note also that motions in the sun's upper layers do not reduce to differential rotation, random complex motions of a smaller scale being added to it, which makes the evolution of the magnetic field correspondingly more complex.

Clearly, our estimate of the magnetic Reynolds number above rests on an a-priori estimate of the scale L of field gradients, or, in more physical terms, on an a-priori estimate of the electric current density which is estimated to be of order $|j| \approx B/(\mu_0 L)$. But the scale L can vary from point to point in the plasma, and, at least in the vicinity of some particular points, it may be much less than the global size of the plasma system involved in the considered motions. This would happen if the electric current circuit is, in some parts at least, unusually strongly concentrated and flows through narrow sections. Making a realistic a-priori estimate of the actual value of the magnetic Reynolds number is then a rather subtle and difficult undertaking. One should check whether the plasma motion indeed does not spontaneously develop such

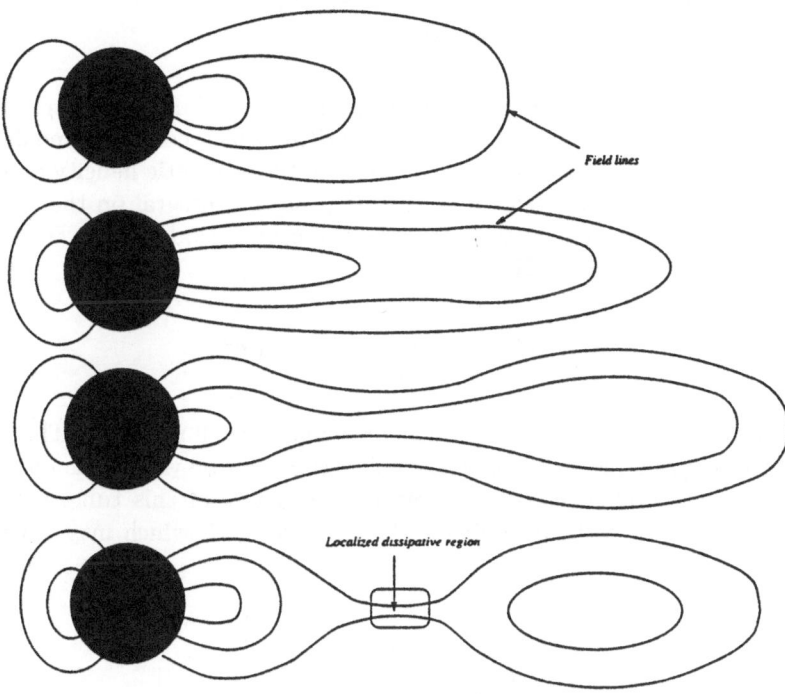

Fig. 3. Evolution under flux-freezing conditions of a field line immersed in the differentially rotating part of the solar interior.

regions of concentrated currents, a necessary condition for maintaining a perfect MHD regime everywhere. The physical effects of a local breakdown of the perfect MHD approximation in restricted regions are described later in this lecture and in the accompanying one by Clare Parnell. In the presence of complex plasma motions, the electric current structure which develops may conceivably also become itself very complex and develop regions of strong concentration, with large magnetic field changes in intensity and direction on rather small scales. Then the question arises as to whether the magnetic Reynolds number calculated for realistic gradient scales remains indeed large, and to define the properties of the regions of space where it does not. Probably, the solar corona, in regions where the magnetic field closes back on the photosphere, is not in a state where perfect MHD is everywhere applicable, due to the permanent growth of complexity induced by boundary fluid motions. This may be the driving reason for its heating, since the growth of small scales promotes Joule dissipation.

15.4   Conservation of Magnetic Helicity
in Perfect MHD Motions

The topological constraints which bear on perfect MHD motions result in conservation of a number of integral quantities which contain some degree of information on this topology, as for example the magnetic helicity of a closed flux tube. Let us define this quantity, H_m, as the integral on the volume of the tube, which may change in time, of the scalar product $(\boldsymbol{A} \cdot \boldsymbol{B})$ where $\boldsymbol{A}$ is a vector potential for $\boldsymbol{B}$.

$$H_m(t) = \int_{tube(t)} (\boldsymbol{A} \cdot \boldsymbol{B}) \, d^3r \qquad (108)$$

The tube is the union of a bunch of elementary flux tubes. The volume element, d^3r, in one such elementary flux tube can be written as $dldS$ where dl is a the line element on the "central" field line of this tube, C, and dS its cross section at the curvilinear absissa l along C, which may vary with l (Fig. 4).

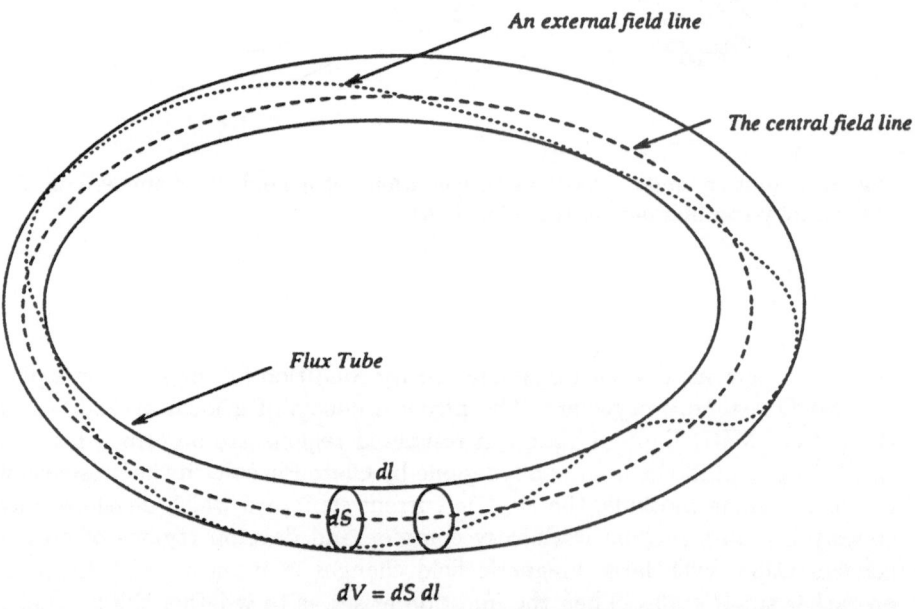

Fig. 4. A closed elementary flux tube is represented. The volume element at some position along its central line is $dV = dS \, dl$, dS being its section and dl the line element along its central field line.

Then let us write that $\boldsymbol{B} = B\boldsymbol{t}$ where $\boldsymbol{t}$ is a unit vector tangent to C while B is the modulus of the field. Then the integral that defines the magnetic helicity of an elementary flux tube can be written as:

$$H_m = \int_C (\boldsymbol{A} \cdot \boldsymbol{t}) \, B \, dS(l) \, dl \tag{109}$$

But $B \, dS(l)$ is the magnetic flux $d\Phi_\perp$ through the elementary tube which is independent of curvilinear abcissa along C, by the very definition of a flux tube. Moreover $(\boldsymbol{A} \cdot \boldsymbol{t}) \, dl$ is the circulation element $(\boldsymbol{A} \cdot \boldsymbol{dl})$ of $\boldsymbol{A}$ along C. Hence

$$H_m = d\Phi_\perp \int_C \boldsymbol{A} \cdot \boldsymbol{dl} \tag{110}$$

The integral along circuit C can be transformed by Stokes flux/circulation theorem in an integral on a surface S spanning C of the rotational of that vector. Here, this results in

$$\int_C \boldsymbol{A} \cdot \boldsymbol{dl} = \int_S \mathrm{rot}\boldsymbol{A} \cdot \boldsymbol{dS} = \int_S \boldsymbol{B} \cdot \boldsymbol{dS} = \Phi(C) \tag{111}$$

where $\Phi(C)$ is the magnetic flux through C. In brief,

$$H_m = d\Phi_\perp \ \Phi(C) \tag{112}$$

In a perfect MHD motion the line C moves conserving the flux $\Phi(C)$ through any surface that it spans, and the tube itself evolves in such a way that it conserves the flux $d\Phi_\perp$ through its section (first flux-freezing theorem). As a result H_m is constant following the motion of an infinitesimal flux tube, provided it closes on itself (Woltjer, 1958). The result extends to a flux tube made of the union of such elementary tubes. Extensions of the concept of magnetic helicity can be defined which apply to portions of such tubes limited by crossing surfaces (Berger et Field, 1984). In the case of closed flux tubes, the above expression makes it immediately apparent that the simple definition of magnetic helicity given above is gauge-invariant. This question was pending since vector potentials $\boldsymbol{A}$ are defined up to a gradient.

15.5   The Concept of Magnetic Reconnection

In which conditions do these perfect MHD constraints cease to apply? It is not enough, in practice, that the magnetic Reynolds number be non-infinite for dissipative effects to become an observable reality. If it is very large, the gradient scale of the magnetic field being L, little deviation from perfect MHD will be apparent until a time

$$\tau_{evol} = \mu_0 \sigma_e L^2 \tag{113}$$

has elapsed, which, for L of order of the size of an astonomical object, is extremely long indeed. The realistic exception to perfect MHD is when the plasma locally develops very small scalelengths of the magnetic field as compared to the global scale of the system in which these MHD motions are produced. But it can be said that such an exception is not exceptional at all. This is because perfect MHD systems, by their own dynamics, often spontaneously develop such regions of strong field gradients. In these localized regions perfect MHD breaks down and dissipative MHD motions develop locally, in an ambient medium where nevertheless perfect MHD still remains valid at large. The local dissipative motion has to somehow match to a general flow which, at other places, retains perfect MHD character (Fig. 5).

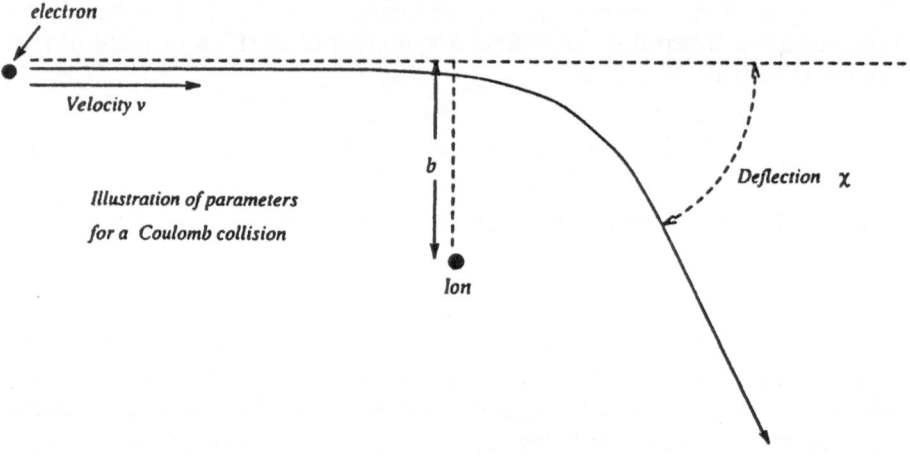

Fig. 5. A simple representation of a reconnection event in the vicinity of some celestial body. The atmosphere is assumed to expand in time to the right. The magnetic structure expands as well, still satisfying, in the first three frames, the topological constraints of perfect MHD. In the fourth frame, a region, indicated by the arrow and the box, where the electric current is concentrated and electrical resistivity influences the plasma motion on a short enough time scale, has appeared. The field lines have been broken, and a reconnection event has started.

Locally, in the dissipative region, however the topological constraints of perfect MHD may be violated. For example, matter can slip there accross field lines. A consequence can be that fluid elements which were initially not magnetically connected would become so after plasma elements to which they were connected happened to flow through such a dissipative region. The description of such motions, which are dissipative in restricted regions of space but otherwise are perfect MHD, is the purpose of "reconnection theory" described by Clare Parnell in this series of lecture.

Part II: How Reliable Is MHD?

16   From a Microscopic to a Macroscopic View of a Plasma

17   What a Plasma Really Is

A plasma is a conducting fluid, gaseous or sometimes liquid. Plasmas are met most frequently in astrophysical or planetological environments. The dense fluids inside stars, sometimes quantum-degenerate, are one example, as are thermally partially ionized plasmas of stellar atmospheres, or gases which are ionized by an external agent, such as the UV radiation of a nearby star in HII nebulae or of the sun in the upper terrestrial ionosphere, or even the cosmic rays in the molecular clouds of the interstellar medium. In the latter case, for example, the ionized fraction would be very low, but such media, eventhough they are poor conductors, still have large magnetic Reynolds numbers because of their size. Given that diversity, one could question whether a conducting fluid dynamic description is always appropriate.

18   What Happens at the Microscopic Level in a Plasma?

A plasma is basically a collection of particles, some or all of them being electrically charged. They are subject to a number of forces, external, that is generated by sources external to the plasma itself, or internal, i.e. created by the particles of the plasma itself. These forces may, for example, be gravitational or electromagnetic, such as the Lorentz force:

$$F = q(E + v \times B) \tag{114}$$

Here v is the velocity of an individual particle, not the velocity of a piece of fluid. Any pair of charged plasma particles interact by the Lorentz force associated to the field that the other create. Let an index i label the different particles. The electric charge of particle i is distributed in space like a Dirac distribution:

$$\rho_e(r, t) = q_i\, \delta(r - r_i(t)) \tag{115}$$

and so is its electric current density:

$$j(r, t) = q_i v_i(t)\, \delta(r - r_i(t)) \tag{116}$$

The associated electric and magnetic forces decrease in proportion to the inverse square of distance and are therefore long-ranged. Interaction forces between neutral particles fall off much faster. As a result, any one given charged particle usually interacts simultaneously with a large number of others. Charged particles may suffer also other interactions, for example with neutral ones or with photons.

19   How to Describe the Dynamics of a Plasma?

What is the outcome of so many interactions among so many particles? Our knowledge of the state of such a big system will always be necessarily incomplete, since it is impossible to keep track of the motion of all particles in a system that may gather, typically, an Avogadro number of them, of order 10^{23}, sometimes much more. A statistical description, the sophistication of which may vary according to our needs, is unavoidable. We shall see that, for times long compared to the characteristic time between collisions and for systems large compared to the mean free path, this description reduces to MHD. However, both the collision frequency and the mean free path vary with the actual physical conditions. Sometimes the system, if its behaviour has to be described on short enough time or spatial scales, is in fact far from the conditions which allow an MHD description. A purely kinetic description is then appropriate.

20   A Methodic Approach to Kinetic Plasma Theory

20.1   The Large Phase Space and the Large Distribution Function

We define a microstate of the plasma as one in which all the positions and velocities of all plasma particles, of which there is a number N, are known with infinite accuracy. To keep simple we disregard quantum indeterminacy and assume that a classical description of the system is justified. The microstate can be thought of as represented by a point $\mathcal{M}$ in a space Ω, the "space of states" or "large space", with $6N$ dimensions, since for each particle we need to know the 3 components of its position vector and the 3 components of its velocity or of its momentum. The microstate $\mathcal{M}$ cannot be known. We therefore define the probability density $\mathcal{F}(\mathcal{M}, t)$ to find it about some state $\mathcal{M}$ of Ω.

20.2   Evolution of $\mathcal{F}$ and the Liouville Equation

When the system evolves in time, the set of states $\mathcal{M}$ that were at time t_1 in a vicinity $d\mathcal{M}_1$ of a state $\mathcal{M}_1$ move at time t_2 in a vicinity $d\mathcal{M}_2$ of a point $\mathcal{M}_2$ of the large space. These microstates transport with them their probability of being actually realized. Hence the probability of finding the microstate in $d\mathcal{M}_1$ at time t_1 is the same as that of finding it at time t_2 in the element $d\mathcal{M}_2$ corresponding to $d\mathcal{M}_1$ by the N-body motion. This is expressed by

$$\mathcal{F}(\mathcal{M}_1, t_1)d\mathcal{M}_1 = \mathcal{F}(\mathcal{M}_2, t_2)d\mathcal{M}_2 \tag{117}$$

A theorem of hamiltonian mechanics, the Liouville theorem, states that the volume element in the large space, more precisely the volume element of the large space of canonical conjugate variables for the N particles, is constant following this element in its N-body motion. In the present case, these variables may be taken as the positions of the particles r_i and their canonical conjugate momenta $P_i = m_i v_i - q_i A(r_i)$ where $A(r)$ is a vector potential of the magnetic field. The hamiltonian function of the system is:

$$\mathcal{H}(r_i, P_i) = \sum_{i=1}^{N} \frac{(P_i + q_i A(r_i))^2}{2m_i} + \frac{1}{2} \sum_{i=1}^{N} \sum_{j \neq i} \frac{q_i q_j}{4\pi\epsilon_0 |r_i - r_j|} \qquad (118)$$

The Liouville theorem states that

$$\left(\prod_{i=1}^{N} d^3 r_i d^3 P_i \right)_{\mathcal{M}_1, t_1} = \left(\prod_{i=1}^{N} d^3 r_i d^3 P_i \right)_{\mathcal{M}_2, t_2} \qquad (119)$$

but, given the relation which exists between the canonical variable P_i of particle i and its position and momentum, we find, calculating the Jacobian of the transformation, that

$$\prod_{i=1}^{N} d^3 r_i \, d^3 P_i = \prod_{i=1}^{N} d^3 r_i \, d^3 p_i = \prod_{i=1}^{N} m_i^3 \, d^3 r_i \, d^3 v_i \qquad (120)$$

All these different volume elements are then conserved following the N-body motion as well. Let us sum up by saying that the Liouville theorem states that

$$d\mathcal{M}_1 = d\mathcal{M}_2 \qquad (121)$$

so that the conservation of probability following the N-body motion reduces to the statement that the large distribution function $\mathcal{F}$ is conserved following this motion:

$$\mathcal{F}(\mathcal{M}_1, t_1) = \mathcal{F}(\mathcal{M}_2, t_2) \qquad (122)$$

Making the time derivative following the N-body motion explicit, the equation $D\mathcal{F}/Dt = 0$ can be written as

$$\frac{\partial \mathcal{F}}{\partial t} + \sum_{x,y,z} \sum_{i=1}^{N} \frac{dx_i}{dt} \frac{\partial \mathcal{F}}{\partial x_i} + \sum_{x,y,z} \sum_{i=1}^{N} \frac{dp_{x,i}}{dt} \frac{\partial \mathcal{F}}{\partial p_{x,i}} = 0 \qquad (123)$$

This is the Liouville equation, which can be written, defining ∂_i as the gradient operator with respect to the i-th particle's momentum p_i and F_i as the force exerted on it, as

$$\frac{\partial \mathcal{F}}{\partial t} + \sum_{i=1}^{N} (v_i \cdot \nabla_i) \mathcal{F} + \sum_{i=1}^{N} (F_i \cdot \partial_i) \mathcal{F} = 0 \qquad (124)$$

This beautiful result gives an *exact* evolution equation for the large distribution function, which is unfortunately useless because the function depends on too large a number of variables. It is necessary to consider so-called reduced distribution functions.

20.3   Reduced Distribution Functions

The one-body reduced distribution function is defined by

$$f_1(\boldsymbol{r}_1, \boldsymbol{p}_1, t) = N \int d2 \int d3.... \int dN \ \mathcal{F}(1, 2, 3, ..., N, t) \qquad (125)$$

The notation 1 represents $(\boldsymbol{r}_1, \boldsymbol{p}_1)$. The notation $d1$ represents $d^3r_1 d^3p_1$. We assume for simplicity that all particles are of the same nature and in practice indistinguishable from eachother. As a result $\mathcal{F}$ is a symmetric function of its arguments 1, 2, ..N.., so that there is actually a unique one-body distribution function (for particles of the same species). The two-body distribution function is defined as

$$f_2(\boldsymbol{r}_1, \boldsymbol{p}_1, \boldsymbol{r}_2, \boldsymbol{p}_2, t) = N(N-1) \int d3.... \int dN \ \mathcal{F}(1, 2, 3, ..., N, t) \qquad (126)$$

For the same symmetry reasons, there is a unique two-body distribution function (among similar pairs of particles). Note that these functions are not probability densities. Because of the factors N and $N(N-1)$, their integrals are not normalized to unity, contrary to $\mathcal{F}$. Their physical interpretation is however simple: $f_1(\boldsymbol{r}, \boldsymbol{p})d^3r d^3p$ is the average value of the number of particles which, at time t, are in $d^3r d^3p$ and $f_2(\boldsymbol{r}, \boldsymbol{p}, \boldsymbol{r}', \boldsymbol{p}')d^3r \ d^3p \ d^3r' \ d^3p'$ is the average number of pairs of particles, one of which is in $d^3r d^3p$ and the other in $d^3r' d^3p'$. One can define similarly 3-body, 4-body, .. distribution functions.

20.4   BBGKY Hierarchy

The equations which describe the time evolution of reduced distribution functions are obtained by acting on the Liouville equation with the same integration operator which defines them in terms of $\mathcal{F}$. This produces for the one-body distribution function $f_1(\boldsymbol{r}_1, \boldsymbol{p}_1, t)$ the evolution equation below, where the time variable has been omitted from the arguments for brevity. $\boldsymbol{F}_{1,ext}(1)$ is the external force exerted on a particle at $\boldsymbol{r}_1$ with momentum $\boldsymbol{p}_1$. The force exerted by a particle 2 at $\boldsymbol{r}_2$ with momentum $\boldsymbol{p}_2$ on a particle 1 at $\boldsymbol{r}_1$ with momentum $\boldsymbol{p}_1$ is noted as $\boldsymbol{F}_{2/1}(\boldsymbol{r}_2, \boldsymbol{p}_2, \boldsymbol{r}_1, \boldsymbol{p}_1)$. The equation for f_1 is

$$\frac{\partial f_1(1)}{\partial t} + (\boldsymbol{v}_1 \cdot \boldsymbol{\nabla}_1)f_1(1) + \boldsymbol{F}_{1,ext} \cdot \boldsymbol{\partial}_1 f_1(1) + \int d2 \ \boldsymbol{F}_{2/1} \cdot \boldsymbol{\partial}_1 f_2(1, 2) = 0 \qquad (127)$$

It is not independent since it involves the two-body distribution function. It must therefore be complemented by an equation which describes the evolution of the function $f_2(1, 2)$ which is similarly obtained and can be written as:

$$\frac{\partial f_2(1, 2)}{\partial t} + (\boldsymbol{v}_1 \cdot \boldsymbol{\nabla}_1 + \boldsymbol{v}_2 \cdot \boldsymbol{\nabla}_2)f_2(1, 2) + (\boldsymbol{F}_{1,ext} \cdot \boldsymbol{\partial}_1 + \boldsymbol{F}_{2,ext} \cdot \boldsymbol{\partial}_2)f_2(1, 2)$$

$$+ \boldsymbol{F}_{2/1} \cdot (\boldsymbol{\partial}_2 - \boldsymbol{\partial}_1)f_2(1, 2) + \int d3 \ (\boldsymbol{F}_{3/1} \cdot \boldsymbol{\partial}_1 + \boldsymbol{F}_{3/2} \cdot \boldsymbol{\partial}_2)f_3(1, 2, 3) = 0 \qquad (128)$$

This equation is also not independent because it involves the three-body distribution function. Proceeding that way further, a chain of equations is generated, the so-called BBGKY hierarchy of equations. In full rigour, this hierarchy would eventually bring us back to the Liouville equation.

20.5 Closure Assumptions and Kinetic Equations for the One-Body Distribution Function

Here physics comes back, in the form of the art of finding adequate approximations which allow to cut the BBGKY hierarchy at an early level. This is indeeed possible in some important cases, the most classical one being the case of a dilute gas, of density n, the particles of which interact by a short-range force, of characteristic interaction distance a. Such a gas is definitely not a plasma! Under such conditions the integral on the states of particle 3 in the equation for f_2 is of order $na^3 \ll 1$ as compared to other terms. Neglecting it, the equation for f_2 can be solved assuming that at time $t = -\infty$ $f_2(1,2) = f_1(1)f_1(2)$ and that f_1 evolves slowly as compared to f_2. The result for f_2 in terms of f_1 is substituted in the first equation of the BBGKY hierarchy. An independent equation for f_1 known as the Boltzmann equation is thus obtained (Uelenbeck et Ford, 1961). This equation for rare, short-range, binary collisions has been the first example ever obtained of a kinetic equation for the one-body distribution function. Ludwig Boltzmann, who first derived it, obtained it by more direct arguments (Huang, 1963), namely by counting how many particles in the mean enter and leave the different cells of the (r, p) space per second due to both collisions and regular forces. Boltzmann's heuristic method can be generalized to a wealth of different situations, like emission and absorption of photons by atoms, or photon scattering by atoms. Radiation transfer equations are obtained that way. They are just kinetic equations for photons coupled to emitting, absorbing and scattering particles. In the case of plasmas, however, one usually is in the limit opposite to that of dilute neutral gases: each particle experiences at one given time interaction with a great many others whereas in the Boltzmann regime interactions are supposedly rare and binary.

21 Strongly and Weakly Coupled Plasmas

A measure of the average intensity of the interaction among particles is the dimensionless number Γ defined as being the ratio of interaction energy at average distance to the average kinetic energy. The average interparticular distance d in a medium of density n is defined by

$$4\pi n d^3/3 = 1 \tag{129}$$

For two ions of charge Ze interacting electrostatically, we then have

$$\Gamma = \left(\frac{4\pi}{3}\right)^{\frac{1}{3}} \frac{Z^2 e^2 n^{\frac{1}{3}}}{4\pi\epsilon_0 k_B T} \tag{130}$$

According to temperature and density the number Γ can be large or small. In tenuous hot plasmas it is generally small. Such plasmas are said to be weakly coupled. Γ takes values larger than unity or even large values in very dense "relatively" cold plasmas like those of the interiors of white dwarf stars or planets. Such plasmas are said to be strongly coupled (see for example Kalman, 1978). In weakly coupled plasmas, particles passing near eachother at the average interparticular distance suffer only weak deviations, because their kinetic energy is much in excess of the interaction energy. The fact that a particle is present at some point r is then but weakly influenced by the presence (or not) of a particle at a point r' distant by an interparticular distance or more: positional correlations are weak at such distances, as would also be velocity correlations. Weak, but not absent. The existence of non-zero position correlations manifests itself in the "screening effect" which can be schematically described as follows. An ion tends to attract electrons and to repell other ions. In some statistical sense, nearby electrons will make a little detour in their unperturbed trajectories to pass closer to the ions, while ions will detour to avoid coming close. There is then be a slight excess of negative charges about any given ion. This negative charge is not made of a bunch of electrons bound to the ion, but of a dynamic grouping in which each individual electron resides only for a short time. To calculate this screening effect simply, the following simple model is useful. The plasma is described as the mixture of an electronic and an ionic fluid. The latter forms an uniform unperturbable background, this being justified by the high ion inertia. In the absence of any perturbation it is assumed that the electron fluid is also uniform and that there is charge neutrality. Let the density of both types of particles in this uniform state, assuming ions to be singly charged, be n_0. A test-ion, a proton, regarded as a positive point charge q_e is introduced at rest at position $r = 0$. It is asked how the electron fluid redistributes itself about the test-ion, reaching a density distribution of spherical symmetry $n(r)$ under the influence of its own electrostatic field, of the electrostatic field of the ion background and of perturbed electron fluid and of the electron pressure force. For simplicity, the electron gas is regarded as an isothermal perfect gas, of temperature T. The electric field in the plasma is given in terms of the electronic density $n_e(r)$ by Poisson's equation. Solving self-consistently this equation and the equation of force equilibrium on the electronic fluid, it is finally found that the electric potential $\Phi(r)$ about the test ion establishes to:

$$\Phi(r) = \frac{q_e}{4\pi\epsilon_0 r} \, e^{-r/\lambda_D} \tag{131}$$

where λ_D, the Debye length, is given by:

$$\lambda_D^2 = \frac{\epsilon_0 k_B T}{n_0 q_e^2} \tag{132}$$

The density of the electrons is only weakly perturbed, because it is found that

$$\frac{n(r) - n_0}{n_0} = \frac{1}{n\lambda_D^3} \left(\frac{\lambda_D}{4\pi r} \, e^{-r/\lambda_D} \right) \tag{133}$$

and, as shown below, our description makes sense only when $n\lambda_D^3 \gg 1$. Comparing to what would be the potential about the test ion in vacuo, we observe an exponential attenuation on the scale of the Debye length. This is due to the gathering of electrons described above. Each individual electron remains at Debye distance from the test ion only for a time $\tau_p = \lambda_D/v_{Te}$, which is very short and can be expressed by:

$$\tau_p^{-2} \equiv \omega_p^2 = \frac{n_0 q_e^2}{\epsilon_0 m_e} \tag{134}$$

The pulsation ω_p is the (electronic) plasma pulsation. It makes sense to represent the electrons as a fluid only if there is a large number of them in the potential structure, of size the Debye length, which they form. A plasma where the number of electrons in the Debye sphere, N_D, is large, that is in which

$$N_D \equiv \frac{4}{3}\pi \, n_0 \lambda_D^3 \gg 1 \tag{135}$$

is said to be a collective plasma, because many particles participate to building up the electric field at any given point. Collective plasmas are also wealy coupled, because Γ and $n_0\lambda_D^3$ are related by

$$\Gamma = \frac{1}{48\pi^2} \frac{1}{(n_0\lambda_D^3)^{\frac{2}{3}}} \tag{136}$$

To sum up, collective plasmas, where $n_0\lambda_D^3$ is large, are also weakly coupled and particles in them are subject to weak positional correlations.

22   Vlassov Dynamics of Collective Plasmas

The weakness of correlations can be taken advantage of for cutting the BBGKY hierarchy. Let us make the correlation part of the two-body distribution function explicit by writing it as

$$f_2(1,2) = f_1(1)f_1(2) + g_2(1,2) \tag{137}$$

where g_2 is the particle correlation function, which is small for collective plasmas. If we simply neglect it we obtain an independent equation for the one-body distribution function:

$$\frac{\partial f_1(1)}{\partial t} + (v_1 \cdot \nabla_1)f_1(1) + (F_{1,ext} + F_{1,av}) \cdot \partial_1 f_1(1) = 0 \tag{138}$$

where the average collective force exerted on particle 1, $F_{1,av}$, is defined at this approximation as the average value of $F_{2/1}$ by the one-body distribution function and not by the two-body distribution function, as should rigourously be and as was the case in the original form of the first equation of the hierarchy. Then, at the "Vlassov approximation",

$$F_{1,av} = \int d^3r_2 \int d^3p_2 \ F_{2/1}(r_2, p_2, r_1, p_1) \ f_1(r_2, p_2) \tag{139}$$

The Vlassov equation so obtained looks like a Liouville equation for a system made of only one particle, but this is only apparent because the average collective force on particle 1, $F_{1,av}$, depends on the unknown of the problem, f_1. Because of that the Vlasov equation is non-linear. It can be given many different equivalent forms. For example if the interaction is only electrostatic, $F_{1,av}$ is but an electric force that can be written as $F_{1,av} = q_1 E_{av}(r_1)$. By definition, this field is created by particles distributed according to the function f_1. The corresponding density of particles at a given point r is then

$$n(r) = \int d^3p \ f_1(r, p) \tag{140}$$

and the associated charge density is

$$\rho_e(r) = q_e \int d^3p \ f_1(r, p) \tag{141}$$

The collective electric field is then given at the Vlassov approximation by Poisson's equation for particles distributed with that charge density:

$$\epsilon_0 \ \mathrm{div} E_{av} = q_e \int d^3p \ f_1(r, p) \tag{142}$$

If there is more than one type of charge carriers, labeled by an index α, with charge q_α, each can be described by its own one-body distribution function $f_{1,\alpha}$ and the collective Vlassov electric field then obeys the Poisson equation

$$\epsilon_0 \ \mathrm{div} E_{av} = \sum_\alpha q_\alpha \int d^3p \ f_{1,\alpha}(r, p) \tag{143}$$

and Vlassov equation for $f_{1,\alpha}$ can be written as:

$$\partial f_{1,\alpha}/\partial t + (v \cdot \nabla) f_{1,\alpha} + (F_{\alpha, ext}(r, p) + q_\alpha E_{av}(r)) \cdot \partial f_{1,\alpha} = 0 \tag{144}$$

It is coupled to the distribution functions of all charged species because E_{av} depends on them all.

23   What Is Lost by Neglecting Correlations?

It can be shown that the Vlassov equation does not drive the plasma towards thermodynamic equilibriuum. Actually it leaves constant the statistical entropy of the system

$$S_{stat} = - \int f_1(r,p) \; log(f_1(r,p)) \; d^3r d^3p \tag{145}$$

It is known, from Boltzmann's famous H-theorem, that the approach to thermodynamic equilibrium implies that this quantity rises to its maximum. By neglecting the correlations, the tendancy towards thermodynamic equilibrium has been lost. This can be understood qualitatively. The Vlassov approximation substitutes to the real particles a distribution by which they are described as if they were smoothly distributed in space with the density $\int f_1(r,p)d^3p$. Therefore random deviations of the particle's motions due to encounters with individually identifiable particles have been completely erased from that description. The electric field is, at this approximation, entirely "regularized", or "smoothed out". But the actual physical reality is still that particles are moving point charges, though certainly in a vast number. The averaged-out electric field described by the Vlassov equation is not the real electric field which exists in the plasma, but only a smoothed-out approximation, though a very good one. The real field differs from this average in that it exhibits fluctuations about this mean due to the discrete character of charged particles. These fluctuations are weak, rapidly varying, but exist and are superposed on the Vlassov field as a noise, the "discrete particle noise" (Fig. 6).

Each fluctuation is the signature of a nearby point charge, the electric signal of which is individualized on the background of the sum of the fields of all the more distant others, which is well represented by the average Vlassov field. These more nearby particles, the electric field of which is felt more personnally, are the ones which produce positional correlations among particles, for they are the ones the field of which is individualized enough to influence, though only weakly, the motion of the neighbouring particle which happen to feel this fluctuation. The effect of these fluctuations constitutes the so called "collisions". The influence of these more individualized encounters will be taken into account by regarding positional correlations as non-zero, thus surpassing the Vlassov approximation. The technicalities are as follows. The correlation function g_2 is regarded as small but non-vanishing. The second equation of the BBGKY hierarchy is linearized with respect to g_2, and the triple correlations, weaker than the second order ones, are neglected. One then solves for g_2 in terms of f_1 (cf for example Ichimaru, 1973). This is the difficult part. The result is inserted in the first equation of the BBGKY hierarchy, and an independent equation for f_1 (simply noted f herafter) is obtained, which can be symbolically written in the following form:

$$\frac{\partial f}{\partial t} + (v \cdot \nabla)f + (F_{ext} + F_{av}) \cdot \partial f = Coll(f) \tag{146}$$

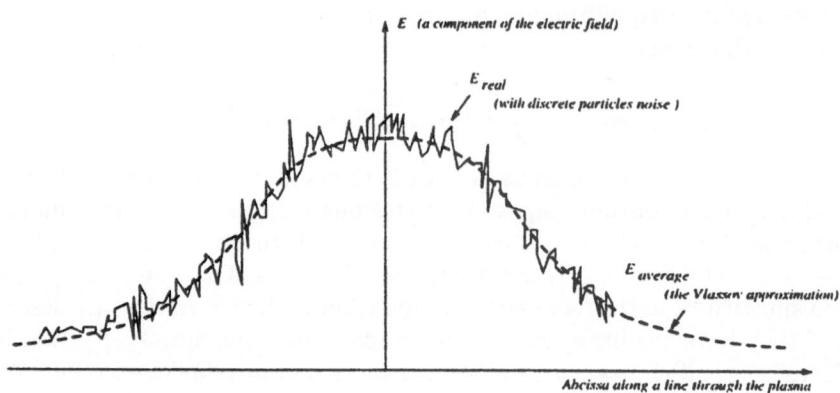

Fig. 6. A schematic sketch of the variation of a component of the electric field along a line through a plasma. The real electric field differs from its Vlassov average, fluctuating about it because of the discrete nature of elementary electrical charges. This "discrete particle noise" causes the real particle dynamics to deviate from the Vlassov one, giving rise to "collisional" effects.

The term $Coll(f)$ represents the effect of the "collisions":

$$Coll(f) = -\int d2 \; \boldsymbol{F}_{2/1} \cdot \boldsymbol{\partial}_1 g_2([f(1), f(2)]) \tag{147}$$

As indicated by the notation, $g_2(1,2)$ can be expressed at this approximation in terms of the one-body distribution function, so that the integral on the right hand side is a functional of it. The collision operator, $Coll(f)$ so obtained defines an evolution equation for f known as the Lenard-Balescu equation (Balescu 1975). The operator $Coll(f)$ is in fact quite complex. It features binary interactions where the colliding particles are screened by a polarization cloud which depends on their velocity and on the value of the one-particle distribution function at that time. In practice more rustic and simple descriptions of the collisions are prefered, which it is not necessary to develop here. It is important however to know that the Lenard Balescu equation, as the simpler Fokker-Planck equation which is often substituted to it, satisfy an H-theorem and cause the statistical entropy to increase, driving the plasma to thermal equilibrium, if of course boundary conditions allow it.

24   Collisional Relaxation Time

When is it possible to use Vlassov equation and when is it needed to take collisional effects into consideration? To answer this question let us calculate how much time it takes collisions to substancially change the trajectory of some particle as compared to what it would have been without them. For that, let us idealize collisions as two-body Coulomb interactions with the nearest neighbour. Consider for example an electron-proton collision. It is a matter of elementary mechanics to calculate the deflexion angle χ in such a collision with impact parameter b and relative velocity v (Fig. 7).

The result is:

$$tg(\frac{\chi}{2}) = \frac{q_e^2}{4\pi\epsilon_0 \ b \ mv^2} \tag{148}$$

Each impact parameter is, for a given velocity associated with a certain deflection $\chi(b)$. The corresponding momentum change in the direction of initial motion is

$$\Delta p_{\parallel} = (1 - cos\chi) \ mv \tag{149}$$

The number of collisions suffered per second by a particle with an impact parameter between b and $b + db$ is:

$$\frac{dN_{coll}}{dt} = 2\pi \ bdb \ nv \tag{150}$$

and the average loss of momentum in the direction of initial motion is :

$$< -\frac{dp_{\parallel}}{dt} >= \int_0^{\infty} mv(1 - cos\chi(b)) \ nv \ 2\pi bdb \tag{151}$$

However, in practice, collisions are not binary ones, but multiparticle ones. This has the effect of screening the interaction at distances larger than the Debye length, λ_D. To take this into account, we should cut the integral above at impact parameters equal or larger than it. Without such a cut, the integral on impact parameters would diverge, a consequence of the long range properties of Coulomb interaction. Then:

$$< -\frac{dp_{\parallel}}{dt} >= nv \int_0^{\lambda_D} mv \ (1 - cos\chi(b)) \ 2\pi bdb \tag{152}$$

and after some elementary algebra using eq. (148):

$$\frac{< -dp_{\parallel}/dt >}{mv} = 4\pi \ n \ v \left(\frac{q_e^2}{4\pi\epsilon_0 mv^2}\right)^2 ln\left(\frac{\lambda_D}{q_e^2/(4\pi\epsilon_0 mv^2)}\right) \tag{153}$$

This relation can be used to define a characteristic collisional deflection time, which turns out to scale with relative velocity as v^3. Faster particles suffer lesser Coulomb friction. Averaging over a maxwellian velocity distribution, we define a thermal collisional frequency for thermal electrons:

$$\nu_{coll} = \frac{3\omega_p}{\sqrt{2}} \frac{ln\Lambda}{\Lambda} \tag{154}$$

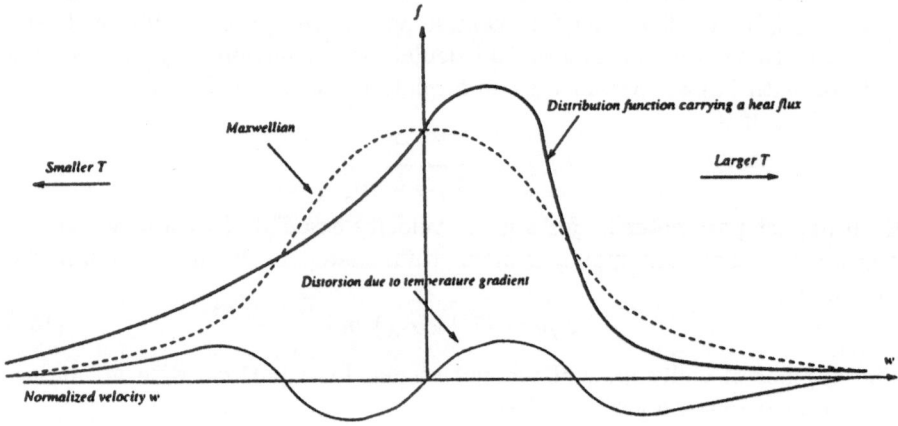

Fig. 7. Illustration of the parameters of a Coulomb collision between an electron and an ion.

where the Coulomb parameter Λ is defined by

$$\Lambda = 12\pi n\lambda_D^3 \tag{155}$$

and its neperian logarithm $ln\Lambda$ is the so-called Coulomb logarithm. In a collective plasma Λ is large. For example it is of order e^{20} in the solar corona. As a result the collisional relaxation time ν_{coll}^{-1} is much longer than the plasma period. The collisional mean free path is $\ell = v\nu_{coll}^{-1}$. Any phenomenon taking place on a time scale shorter than ν_{coll}^{-1} can be described in the framework of the Vlassov collisionless theory. Any phenomenon taking place on a time scale much longer than ν_{coll}^{-1} or on a scale length much larger than ℓ needs a description accounting for collisional effects: in the framework of plasma kinetic theory, the Lenard-Balescu, or perhaps the simpler Fokker-Planck equation should be used. However, this may turn out to be rather tedious. In some cases, the much simpler MHD approximation can be used instead. Before leaving this chapter, the reader should be warned that there exist different types of collisional relaxation times, which can largely differ because the electron to ion mass ratio m_e/m_i is very small (Spitzer, 1962). As a result collisions between electrons and ions are almost elastic and energy exchanges between these two populations take place on a time scale longer than deflection effects. Moreover the decrease of the collision frequency with velocity as $1/v^3$, a property of Coulomb collisions, has important consequences too. Once accelerated to suprathermal velocities, it becomes rather difficult for fast particles to slow down and thermalize. If suprathermal enough they re-

main as a collisionless population for rather long times, their behaviour being
described by Vlassov dynamics, in a thermal particle background that is itself
a lot more collisional.

25   The Hydrodynamic Limit

The hydrodynamic limit is met when the scale length over which plasma
properties, such as for example the root mean square velocity of plasma particle about their mean value, becomes large as compared to the mean free
path ℓ. and the time scale for these variations is large as compared to the collisional time scale ν_{coll}^{-1}. Then, in the vicinity of some point r in the plasma,
particles undergo one and their next collision under almost identical envi-
ronmental conditions. At some rough degree of approximation, one could say
that collisional relaxation takes place as if particles were in an homogeneous
and stationary medium with the local properties. The outcome of such an
interaction process is well known: after a few collision times the medium
reaches a state of thermodynamic equilibrium. The one particle distribution
function becomes very close to a maxwellian in the fluid's local frame of rest.
In the laboratory frame, the fluid may have a global translation velocity and
its one particle distribution function is:

$$f(r,v,t) = \frac{n(r,t)}{(2\pi k_B T(r,t)\,/m)^{\frac{3}{2}}} \; exp\left(-\frac{1}{2}\frac{m(v - u(r,t))^2}{k_B T(r,t)}\right) \qquad (156)$$

In fact such an equilibrium is reached only in a local sense. It is approximately as described by equation (156) only in a small vicinity of point r and
for some short time lapse. This vicinity is nevertheless large compared to the
mean free path ℓ, but small compared to the gradient scale L of macroscopic
plasma properties, such as density n, or temperature T or fluid velocity u.
Similarly the time scale over which the plasma is well described by this distribution is long compared to the collisional relaxation time scale, but short
as compared to the characteristic time scale for the evolution of macroscopic
quantities. The latter remain dependent on position r and time t, but vary
over scales longer than ℓ and ν_{coll}^{-1} resp. The problem therefore reduces in
this limit to calculating the three functions $n(r,t)$, $T(r,t)$ and $u(r,t)$, which
is a vastly simpler undertaking than calculating a general distribution function $f(v,r,t)$ for all values of v ! It will be seen that the evolution of these
functions is actually ruled by the equations of hydrodynamics or MHD. So
doing, we shall discover that the distribution function above is itself but a
first approximation to the real distribution function. We will have to calculate a better approximation for accounting for the effects of inhomogeneity
which give rise to transport processes.

26   Transport Processes

As mentioned, the maxwellian shape of the distribution function is but a first approximation, valid when the inhomogeneity of macroscopic quantities can be totally ignored, in which case the successive collisions take place in strictly identical environments. But this would be so only if the mean free path ℓ were strictly zero. In reality ℓ is only very small, in the sense that $\ell \ll L$, but not zero. The ratio of the mean free path to the gradient scale, $\mathcal{K} = \ell/L$, is the Knudsen number. It is small but non-zero and can therefore be used as a basis for a power series expansion of the distribution function. The zeroth order term of this expansion should be the maxwellian with local values of density, temperature and fluid velocity, for this is what we should get when reducing ℓ strictly to zero. The subsequent terms of the expansion constitute a small departure from maxwellianess which keep track of the existence of actual inhomogeneities in the medium. The first of these extra terms in the expansion in fact constitute a sufficiently good approximation for our needs. It can be calculated by linearization in terms of the small parameter $\mathcal{K}$. This departure from maxwellianess is important, as we shall see, because it carries fluxes of physical quantities to which the isotropic maxwellian distribution contributes for nothing. Hence the calculation of this first order term in the Knudsen number is a must, for without it important physical processes would not be accounted for.

26.1   Heat Conduction in a One-Dimensional Plasma

Let us illustrate how such an expansion can be done on the example of heat conduction in a simplified, one-dimensional, plasma model. This plasma is assumed to suffer no external nor collective forces and collisions are represented in the kinetic equation by a collision operator $\mathcal{C}oll$. We seek for a stationary state. The distribution function $f(x, v)$ must be a solution of the kinetic equation:

$$v \, \frac{\partial f}{\partial x} = \mathcal{C}oll(f) \tag{157}$$

In order of magnitude estimate, the left hand side term is of order $\mathcal{O}(vf/L))$, whereas the right hand side one is of order $\mathcal{O}(f/\tau_{coll})$, τ_{coll} being the collisional relaxation time. Their ratio is of order of the Knudsen number since $v\tau_{coll}/L \approx \ell/L$. To zeroth order in $\mathcal{K}$, the equation then reduces to

$$\mathcal{C}oll(f) = 0 \tag{158}$$

the solution of which is a maxwellian distribution since this equation expresses the fact that collisional relaxation has come to completion, and therefore thermodynamic equilibrium has been reached, a property which results from so-called H-theorems. Disregarding for simplicity a possible fluid motion, the one-dimensional equilibrium distribution function is:

$$f(x, v) = f_M(x, v) = \frac{n(x)}{(2\pi k_B T(x)/m)^{\frac{1}{2}}} \, exp\left(-\frac{1}{2}\frac{mv^2}{k_B T}\right) \tag{159}$$

Let us now proceed to the next term in the $\mathcal{K}$ expansion by writing the real distribution function f as

$$f(x, v) = f_M(x, v) + \mathcal{K}\phi(x, v) \tag{160}$$

We calculate ϕ by perturbation, linearizing the kinetic equation with respect to $\mathcal{K}$:

$$v\frac{\partial}{\partial x}(f_M + \mathcal{K}\phi) = Coll\,(f_M + \mathcal{K}\phi) \tag{161}$$

Since $Coll(f_M)$ vanishes, $Coll(f_M + \mathcal{K}\phi)$ must be the result of some linear operator acting on $\mathcal{K}\phi$ alone. Let us assume that it can be written as:

$$Coll\,(f_M + \mathcal{K}\phi) = -\nu\mathcal{K}\phi \tag{162}$$

where ν is some frequency, of order of the collision frequency, that might depend on x and v. We assume for the sake of a simple illustration of the method that it is a constant. The minus sign on the r.h.s. of equation (162) indicates that, in the absence of inhomogeneities, the temporal variations of f would result in a damping of departures from thermodynamic equilibrium distribution function. Let $\nu^{-1} = \tau$ and let us un-dimensionalyze the variables by scaling x to L, the characteristic gradient scale, and v to a "thermal" velocity v_T. The dimensionless particle velocity w and dimensionless position ξ are:

$$w = \frac{v}{v_T} \qquad\qquad \xi = \frac{x}{L} \tag{163}$$

and the kinetic equation becomes

$$\left(\frac{\tau v_T}{L}\right)\, w\, \frac{\partial}{\partial \xi}(f_M + \mathcal{K}\phi) = -\mathcal{K}\phi \tag{164}$$

Note that $(\tau v_T/L)$ is the Knudsen number. To first order in it, the kinetic equation reduces to

$$-w\,\frac{\partial f_M}{\partial \xi} = \phi \tag{165}$$

which is solved for the first order correction $\mathcal{K}\phi$ to f_M. Restauring dimensional variables, this gives:

$$\mathcal{K}\phi = -\left(\frac{\tau v_T}{L}\right)\, w\, \frac{\partial f_M}{\partial \xi} = -\tau\, v\, \frac{\partial f_M}{\partial x} \tag{166}$$

It is a simple matter to calculate $\partial f_M/\partial x$:

$$\frac{\partial log f_M}{\partial x} = \frac{1}{f_M}\frac{\partial f_M}{\partial x} = \frac{\partial log n}{\partial x} - \frac{1}{2}\frac{\partial log T}{\partial x} + \frac{1}{2}\frac{mv^2}{k_B T}\frac{1}{T}\frac{\partial T}{\partial x} \tag{167}$$

If the medium is a perfect gas in pressure equilibrium, $\partial log n/\partial x + \partial log T/\partial x = 0$. Eliminating the density we find

$$\frac{\partial log f_M}{\partial x} = \frac{1}{2}T\frac{\partial T}{\partial x}\left(\frac{mv^2}{k_BT} - 3\right) \tag{168}$$

whence:

$$\mathcal{K}\phi = -\frac{\tau}{2}\left(\frac{1}{T}\frac{\partial T}{\partial x}\right)v\left(\frac{mv^2}{k_BT} - 3\right)f_M = -\frac{\tau}{2}\left(\frac{1}{T}\frac{\partial T}{\partial x}\right)\frac{n}{\sqrt{2\pi}}\ w(w^2 - 3)e^{-w^2/2}$$
$$\tag{169}$$

This function is represented on Fig. 8 for positive dT/dx together with its sum with $f_M = (n/v_T\sqrt{2\pi})exp(-w^2/2)$.

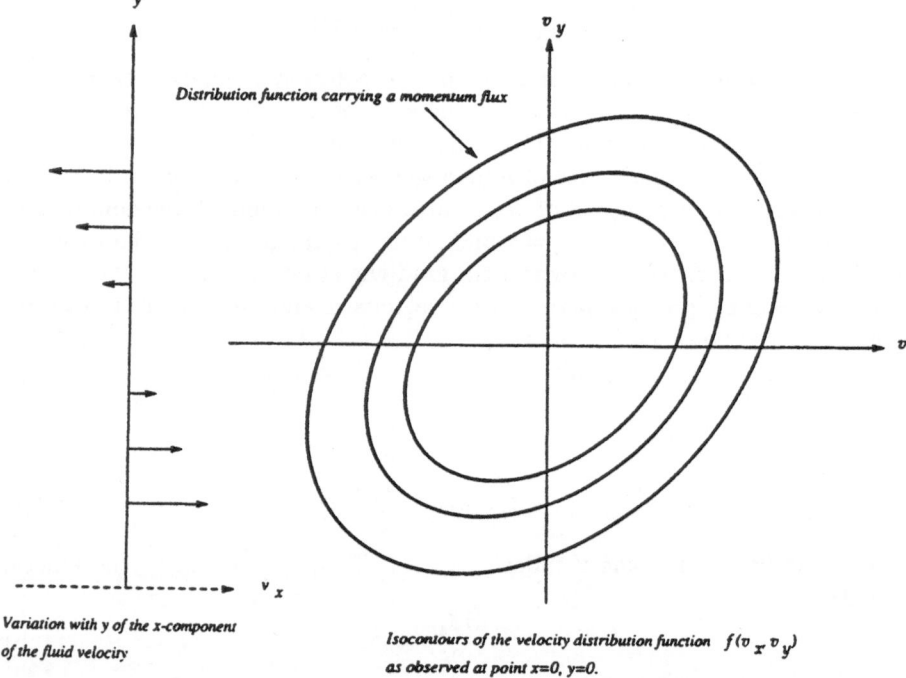

Fig. 8. The distribution function of particles in the presence of a weak temperature gradient. The distribution function is distorted with respect to a perfectly maxwellian shape. The distorsion is responsible for heat conduction, supporting an energy flux opposite to the temperature gradient.

The distribution function which results from this addition of an even and an odd function of v is asymmetrical, or "skewed". This asymmetry, as shown below, is the cause of heat conduction. This result can be qualitatively described as follows. At point x some particles travel to the right and some other travel to the left. The former had their last collision to the left of x, where they have been almost thermalized with a medium that is slightly cooler than at x. As a result these particles are slightly under-energetic in terms of the temperature at x. And conversely for those particles that travel to the left. The heat flux q which is carried by the distribution function (160) is

$$q(x) = \int_{-\infty}^{+\infty} (\frac{1}{2}mv^2)vf(x,v)dv = \int_{-\infty}^{+\infty} \frac{1}{2}mv^3 f_M(x,v)dv + \int_{-\infty}^{+\infty} \frac{1}{2}mv^3 K\phi dv \tag{170}$$

The contribution brought to this integral by f_M is zero for parity reasons. Only the first order term, the distorsion to the maxwellian distribution, makes a contribution to the heat flux, namely, from equation (169):

$$q = \frac{1}{2}mv_T^4 \left(-\frac{\tau}{2}\frac{1}{T}\frac{\partial T}{\partial x} \right) n(x) \int_{-\infty}^{+\infty} w^4(w^2-3)\frac{e^{-\frac{w^2}{2}}}{\sqrt{2\pi}}dw \tag{171}$$

The integral which appears in this expression has a certain, unimportant, numerical value, call it I. We then have obtained the heat flux in the form of a Fourier law:

$$q = -\chi \frac{dT}{dx} \tag{172}$$

Note that q scales proportional to the Knudsen number, since

$$K \approx \frac{\tau v_T}{L} \approx \tau v_T \left(\frac{1}{T}\frac{dT}{dx} \right) \tag{173}$$

and

$$q \approx -\frac{1}{4} KI\; nmv_T^3 \tag{174}$$

Since K is supposedly small, this theory describes heat transport correctly only when

$$q \ll q_{sat} \approx nmv_T^3 \tag{175}$$

If the heat flux in a plasma were to approach this limit, the distorsion of the actual distribution function with respect to the maxwellian would cease to be small and could not be calculated perturbatively.

26.2 Negative Proportionality of Flux to Gradient

More generally, a gradient of any macroscopic quantity has, in the hydrodynamic regime, a microscopic flux associated to it which is caused by the associated distorsion of the actual distribution function with respect to the local thermodynamical equilibrium one. This flux is negatively proportional to the gradient of the macroscopic quantity transported. One form of such a relation is Fourier's law of heat conduction,

$$q = -\chi \nabla T \tag{176}$$

and another is Fick's law of transport by diffusion of a minor species, which expresses the diffusion flux F as:

$$F = -D\nabla n \tag{177}$$

26.3   Anisotropic Transport

In an anisotropic medium, for example one which is strongly enough magnetized for the collision frequency of some particles to be smaller than their gyro-frequencies, the transport coefficients, like χ, D or the electric conductivity σ_e may become tensorial. Specifically, thermal conduction in a plasma is mainly caused by electron transport. In the presence of a magnetic field it becomes strongly anisotropic when their mean free path becomes longer than their gyroradius, because between two successive collisions electrons cannot move farther away cross-fields than a gyroradius, an usually very small distance, whereas they remain free to move any distance along the magnetic field lines. As a result, the heat conductivity is large along field lines but smaller perpendicular to them. The thermal conductivity of a non-magnetized plasma, or its field-aligned part (which gives the field-aligned heat flux as a response to the field-aligned temperature gradient), is given, with T in degrees Kelvin, by:

$$\chi = 1.85 \ 10^{-10} \frac{T^{\frac{5}{2}}}{ln\Lambda} \quad J \ m^{-1} \ s^{-1} \tag{178}$$

The electrical conductivity, which gives the electric charge flux in response to a gradient of electric potential is, in a plasma where the collision frequency of electrons exceeds their gyrofrequency $\omega_{Be} = -q_e B/m_e$:

$$\sigma_e = 1.6 \ 10^{-2} \frac{T^{\frac{3}{2}}}{ln\Lambda} \quad Mho \ m^{-1} \tag{179}$$

By contrast, it becomes very anisotropic when the electron collision frequency is much less than their gyrofrequency. Let us introduce the vector notation $\boldsymbol{\omega}_{Be} = -q_e \boldsymbol{B}/m_e$. The anisotropic electric conduction can easily be calculated adopting a simplified model of plasma dynamics. Assume the ionization to be complete and the ions to be protons. Electrons and protons are treated as two independent fluids coupled by a friction force (Cowling, 1957). The equation of motion of electrons, of charge q_e, mass m_e, density n_e and velocity in the laboratory frame $\boldsymbol{v}_e$ can be written, neglecting gravity:

$$n_e m_e \left(\frac{\partial \boldsymbol{v}_e}{\partial t} + (\boldsymbol{v}_e \cdot \boldsymbol{\nabla})\boldsymbol{v}_e \right) = -\boldsymbol{\nabla} p_e + q_e(\boldsymbol{E} + \boldsymbol{v}_e \times \boldsymbol{B}) - n_e m_e \nu(\boldsymbol{v}_i - \boldsymbol{v}_e) \tag{180}$$

With similar notations, the equation for ions motion is :

$$n_i m_i \left(\frac{\partial v_i}{\partial t} + (v_i \cdot \nabla) v_i\right) = -\nabla p_i + q_i (E + v_i \times B) - n_e m_e \nu (v_e - v_i) \quad (181)$$

The friction on ions must, by action-reaction theorem, be opposite to that on electrons, which is taken into account in equation (181). Let us introduce the differential velocity between electrons and ions, w:

$$v_e = v_i + w \qquad (182)$$

The electric current density j is

$$j = n_i q_i v_i + n_e q_e v_e = n_e q_e (v_e - v_i) = n_e q_e w \qquad (183)$$

The quasi-neutrality of the plasma has been used to write down this expression. Divide the electron motion equation by $n_e m_e$, that of ions by $n_i m_i$ and substract, obtaining an equation for w:

$$\frac{\partial w}{\partial t} + (((v_i + w) \cdot \nabla)(v_i + w)) - (v_i \cdot \nabla) v_i = -\frac{\nabla p_e}{n_e m_e} + \frac{\nabla p_i}{n_i m_i} + \left(\frac{q_e}{m_e} - \frac{q_i}{m_i}\right) E$$

$$+ \frac{q_e}{m_e}(v_i + w) \times B - \frac{q_i}{m_i} v_i \times B - \nu \left(1 + \frac{m_e}{m_i}\right) w \qquad (184)$$

It simplifies in the limit of infinitely heavy ions $m_i = \infty$, and can be converted into an equation for j multiplying by $n_e q_e$, which can be first written as:

$$\left(\frac{\partial}{\partial t}(n_e q_e w) - w \frac{\partial}{\partial t}(n_e q_e)\right) + n_e q_e (v_i \cdot \nabla) w + n_e q_e (w \cdot \nabla) v_i + n_e q_e (w \cdot \nabla) w =$$

$$-\frac{q_e}{m_e} \nabla p_e + \frac{n_e q_e^2}{m_e}(E + v_i \times B) + \frac{q_e}{m_e} j \times B - \nu j \qquad (185)$$

From electron number conservation equation $\partial n_e / \partial t + \mathrm{div}(n_e v_e) = 0$:

$$\frac{\partial j}{\partial t} + w \, \mathrm{div}(n_e q_e (v_i + w)) + n_e q_e (v_i \cdot \nabla) w + (j \cdot \nabla) v_i + (j \cdot \nabla) w =$$

$$\frac{n_e q_e^2}{m_e}\left(E + v_i \times B - \frac{\nabla p_e}{n_e q_e}\right) + \frac{q_e}{m_e} j \times B - \nu j \qquad (186)$$

We can add on the left hand side the term $v_i \mathrm{div} j$, since, for negligible displacement currents, as here, $\mathrm{div} j = 0$. Thus, rearranging notations slightly:

$$\frac{\partial j}{\partial t} + w \, \mathrm{div}(n_e q_e v_i) + (n_e q_e v_i \cdot \nabla) w + w \, \mathrm{div} j + (j \cdot \nabla) w + v_i \, \mathrm{div} j + (j \cdot \nabla) v_i$$

$$= \frac{n_e q_e^2}{m_e}\left(E + v_i \times B - \frac{\nabla p_e}{n_e q_e}\right) + \frac{q_e}{m_e} j \times B - \nu j \qquad (187)$$

The terms on the left hand side can be given an elegant tensor form. Actually, analyzing components, it can be recognized that the following identity holds:

$$(\boldsymbol{a} \cdot \boldsymbol{\nabla})\boldsymbol{b} + \boldsymbol{b} \, \text{div}(\boldsymbol{a}) = \text{div}(\overline{\overline{ab}})$$

whence

$$\frac{\partial \boldsymbol{j}}{\partial t} + \text{div}\left(\overline{\overline{v_i j}} + \overline{\overline{j v_i}} + \frac{\overline{\overline{jj}}}{n_e q_e}\right) + \nu \boldsymbol{j} + \boldsymbol{j} \times \boldsymbol{\omega}_{Be} = \frac{n_e q_e^2}{m_e}\left(\boldsymbol{E} + \boldsymbol{v}_i \times \boldsymbol{B} - \frac{\boldsymbol{\nabla} p_e}{n_e q_e}\right)$$

(188)

This equation gives the electric current in terms of electric and electro-motive fields. It is a generalization of Ohm's law. At this point no use has yet been made of the particular circumstances which make the plasma almost MHD, i.e. gradient scales larger than mean free paths and variation time scales longer than the collision time. In such circumstances, the quantity $\nu \boldsymbol{j}$ is likely to be one of the dominant terms. Let us compare the other terms to it. Clearly $\partial \boldsymbol{j}/\partial t$ is much less because the variation time scale is longer than the collision time. The ratio of the tensorial terms under the div operator are small compared to $\nu \boldsymbol{j}$ because the ratio of these quantities is of order $(jv_i/L)/(\nu j)$, with L the gradient scale, and $(v_i/\nu)/L < (v_{Te}/\nu)/L = \ell/L = \mathcal{K}$. For very collisional systems, ν should exceed ω_{Be}, but this is often not so in reality. Finally the term $q_e \nabla p_e/m_e$ can be compared to νj by noting that the pressure is given, in order of magnitude, by the equation of motion of the fluid as a whole. For a subalfvenic flow, $\nabla p \approx |\boldsymbol{j} \times \boldsymbol{B}|$ and the ratio $(q_e \nabla p_e/m_e)/(\nu j)$ is almost equal to ω_{Be}/ν, which may not be so small. For superalfvenic flow, the estimate is even less favourable. We conclude nevertheless that for $\omega_{Be} \ll \nu$ the electronic pressure term is negligible. In that case, the big equation (188) for $\boldsymbol{j}$ reduces to the usual Ohm's law of MHD:

$$\boldsymbol{j} = \frac{n_e q_e^2}{m_e \nu}\left(\boldsymbol{E} + \boldsymbol{v}_i \times \boldsymbol{B}\right)$$

(189)

If on the other hand ω_{Be} is not much less than ν, the electron pressure term should be retained as an electromotive term and the term $\boldsymbol{j} \times \boldsymbol{\omega}_{Be}$ cannot be neglected as compared to $\nu \boldsymbol{j}$. We leave it to the reader to solve for $\boldsymbol{j}$ by inverting the matrix $\mathcal{N}$ defined by:

$$\mathcal{N}\boldsymbol{j} = \nu \boldsymbol{j} + \boldsymbol{j} \times \boldsymbol{\omega}_{Be}$$

(190)

As $\mathcal{N}$ itself, the inverse matrix is not diagonal, An anisotropic conductivity appears, with so-called direct, Hall and Pedersen components defined by the ratios $(j_\parallel/E_\parallel)$, $(j_\times/E_\perp)$ and $(j_\perp/E_\perp)$.

26.4   Viscosity

The gradients of the components of the bulk fluid velocity, $\boldsymbol{u}$, also create transport effects. The quantitity associated to fluid velocity is momentum.

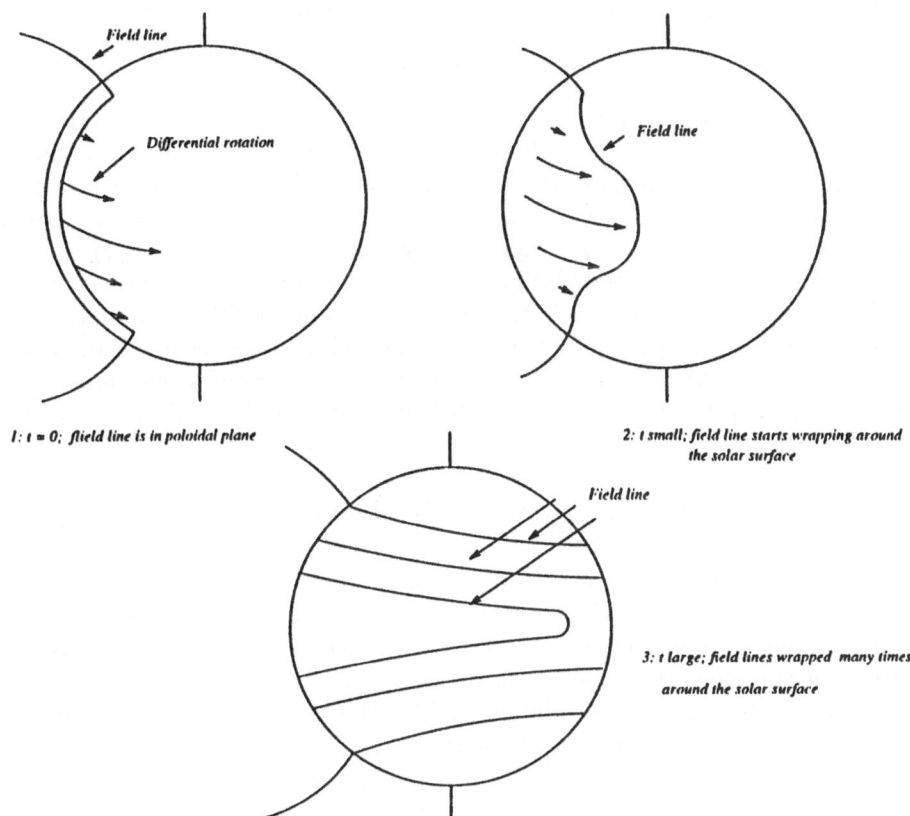

I: $t = 0$; fiield line is in poloidal plane

2: t small; field line starts wrapping around the solar surface

3: t large; field lines wrapped many times around the solar surface

Fig. 9. Isocontours in particle velocity space of the two-dimensional distribution function at position $x = 0$, $y = 0$ in the presence of a weak gradient in the y-direction of a bulk fluid velocity aligned to the x-direction. The bulk velocity profile is represented in the left panel as a function of position in space (fixed x, varying y). The distribution function is distorted with respect to a perfectly maxwellian shape. The distorsion is responsible for viscosity, because it supports a flux of x-component of momentum opposite to its gradient in the y-direction.

Its gradients cause distorsions of the actual distribution function with respect to the maxwellian one which carry a momentum flux, which is just the viscosity phenomenon. Figure 9 shows, in two dimensions, the general shape of a distribution function distorted by a gradient in the y-direction of the x-component of the bulk fluid velocity u. The microscopic flux of momentum is the opposite of the viscous stress tensor, $\overline{\overline{\sigma}}$ presented in Part I.

From the general fact that microscopic fluxes are proportional to the gradients of the transported macroscopic quantity, one infers that the components of $\overline{\overline{\sigma}}$ should depend linearly on the gradients of the components of the bulk velocity $u(r,t)$. In a given rest frame, there should then exist viscosity coefficients A_{ijkl} such that (we use the dummy index rule here):

$$\sigma_{ij} = A_{ijkl}\nabla_k u_l \qquad (191)$$

The A_{ijkl}'s must however satisfy some relations. Indeed, if the medium has isotropic properties, which is often the case for viscosity because it is mainly due to ion contributions, for which the gyrofrequency may indeed be less than the collision frequency, there should be symmetries due to invariance under rotation of both the flux and the velocity field. For example $Axyxy$ should equal $Ayzyz$. There should also be no viscous stresses if the fluid rotates as a solid body, with a velocity field $\mathbf{\Omega} \times r$, whatever the axis $\mathbf{\Omega}$ about which such a rotation is to take place. Taking into account all the constraints so imposed, it can be shown that for an isotropic medium the viscosity coefficients A_{ijkl} can be expressed in terms of two coefficients η and ζ only, so that the viscous stress tensor can be written as:

$$\sigma_{ij} = \eta \left(\nabla_i u_j + \nabla_j u_i - \frac{2}{3}\delta_{ij}\text{div}\mathbf{u} \right) + \zeta\, \delta_{ij}\text{div}\mathbf{u} \qquad (192)$$

We shall not expand here on the reason why the terms proportional to $\text{div}\mathbf{u}$ have been split the way they appear in this expression, nor on the different physical effects represented by each of them (some more details in Heyvaerts 1991). It turns out that the second coefficient of dynamical viscosity, or bulk viscosity coefficient, ζ, is usually negligible. The first coefficient of dynamical viscosity, η, or shear viscosity coefficient, is, in an isotropic plasma, approximately given by

$$\eta \approx 0.2\ 10^{-5}\ \frac{T^{\frac{5}{2}}}{ln\Lambda}\ Kg\ m^{-1}\ s^{-1} \qquad (193)$$

27   From Kinetic Theory to Hydrodynamics

As indicated above, the great simplification of hydrodynamics as compared to full kinetic theory is that the unknowns of hydrodynamics reduce to a limited number of hydrodynamic fields, the density $n(r,t)$, the fluid velocity $u(r,t)$ and the temperature $T(r,t)$. In the small Knudsen number limit, in which hydrodynamics is justified, the kinetic problem is almost completely solved in that the distribution function is expressed as the sum of a main part which is a thermodynamic equilibrium distribution and a small correction to it, $\mathcal{K}\phi$ which is entirely expressible in terms of the gradients of the quantities that define the local thermodynamic equilibrium part. To obtain evolution equations for these macroscopic quantities, it suffices to enter this expression in the kinetic equation and to act on it with an integral operator which singles out the desired macroscopic quantity. For example, integration of the distribution function on velocities gives the density $n(r,t)$, while multiplication by v followed by an integration gives $nu(r,t)$. Acting with such operators on the kinetic equation gives a hierarchy of equations expressing $\partial n/\partial t$, $\partial \boldsymbol{u}/\partial t$, $\partial T/\partial t$, etc.. This is a slightly tedious procedure, in which care should be exerted to take properly into account the different species of particles. We shall not dwell into the details here (some more details in Heyvaerts 1991), but just illustrate the idea on the example of the density. Let $f(\boldsymbol{r}, \boldsymbol{p}, t)$ be the one particle distribution function where the variables are now position and particle momentum. It satisfies the kinetic equation:

$$\frac{\partial f}{\partial t} + (\boldsymbol{v} \cdot \boldsymbol{\nabla})f + (\boldsymbol{F} \cdot \boldsymbol{\partial})f = Coll(f) \tag{194}$$

The particle density at $\boldsymbol{r}$ at time t is

$$n(\boldsymbol{r}, t) = \int f(\boldsymbol{r}, \boldsymbol{p}, t)d^3p \tag{195}$$

Integrate the kinetic equation on momenta, taking into account the simplifying fact that most often the i-component of the force $\boldsymbol{F}$ exerted on a particle does not depend on the same-i component of its velocity. This is true of all forces that do not depend on velocity and also of the Lorentz force. Integrating by parts with f approaching 0 when $\mid \boldsymbol{r} \mid$ and $\mid \boldsymbol{p} \mid$ approach infinity, we obtain:

$$\int \frac{\partial f}{\partial t}d^3p = \frac{\partial n}{\partial t} \tag{196}$$

$$\int (\boldsymbol{v} \cdot \boldsymbol{\nabla})fd^3p = \text{div}(n\boldsymbol{u}) \tag{197}$$

$$\int (\boldsymbol{F} \cdot \boldsymbol{\partial})fd^3p = 0 \tag{198}$$

$$\int Coll(f)d^3p = \dot{n}_{inel} \tag{199}$$

This last term represents the rate of creation of particles of the considered species per unit volume brought about by inelastic collisions, in which particles of this species can be created or destroyed, such as chemical or nuclear reactions, ionizations or recombinations, pair creations etc.. We so obtain a conservation equation for the number of particles of this species:

$$\frac{\partial n_a}{\partial t} + \text{div}(n_a \boldsymbol{u}_a) = \dot{n}_{inel,a} \tag{200}$$

The mass density of the fluid as a whole, ρ, is defined in terms of the number density of the different species a with particle mass m_a and electric charge q_a as

$$\rho = \sum_a n_a m_a \tag{201}$$

while the fluid bulk velocity $\boldsymbol{u}$ is similarly defined in terms of those and of the species bulk velocity by

$$\rho\boldsymbol{u} = \sum_a n_a m_a \boldsymbol{u}_a \tag{202}$$

The equation for ρ is obtained from the species number conservation equations by the operation $\sum_a m_a \times$. In non-relativistic dynamics the total mass is conserved in any type of chemical or ionization reaction, so that

$$\sum_a m_a \, \dot{n}_{inel,a} = 0 \tag{203}$$

We are then left with

$$\frac{\partial \rho}{\partial t} + \text{div}(\rho\boldsymbol{u}) = 0 \tag{204}$$

which is just the mass conservation equation. The electric charge, another conservative quantity, obeys a conservation equation which is obtained similarly by the operation $\sum_a q_a \times$. Had we multiplied the kinetic equation of each species by $\boldsymbol{p}$ before integrating on this variable, we would have obtained a conservation equation for the momentum of that species, and, summing on them, a conservation equation for the momentum of the fluid as a whole. Pre-multiplying the kinetic equation by $p^2/2m_a$ would have given a conservation equation for the kinetic energy (bulk and random) of each species, that could then be globalized by summing on species. All MHD equations can be recovered that way. They have been admittedly obtained in Part I more straightforwardly, but at the expense of a somewhat a-prioristic introduction of hydrodynamical concepts. A slightly more detailed presentation is to be found in this series in Heyvaerts (1991).

28   Couplings Between MHD and Collisionless Phenomena

Again, MHD is valid for plasma motions the scale length of which is larger than all particle's mean free path and the time scale of which is longer than the collisional relaxation time, whereas the Vlassov description applies in the opposite situation. However, the mean free path and the collision frequency depend on the energy of particles, and the gradient scales and evolution time may vary with position and time. So, it is unfortunately not granted that the two situations be mutually exclusive. We expand on these complications below.

28.1   Possible Coexistence of Particles in Hydrodynamic and in Collisionless Regimes

Different populations of particles may coexist in a plasma. For example, occasionnally or permanently, a high energy population may coexist with a lower energy population. These populations will have different collision frequencies and mean free paths. It may happen that the hotter one be in a collisonless, Vlassov, regime while, on similar time and spatial scales, the colder one be in an hydrodynamical regime. Though usually less dense, the hotter component could make a very significant contribution to the electric current or to the heat flux, and could for this reason not be disregarded or treated as an insignificant correction. Such situations are met in the terrestrial magnetosphere, in the interstellar medium, where a cosmic ray population contributes very significantly to the pressure, and occasionnaly in the solar corona, near particle acceleration sites of solar flares. These energetic electron populations probably make the dominant contribution to the heat flux from the heated region to the lower solar atmosphere in the form of beams of precipitating particles, the distribution function of which is certainly far from maxwellian and even far from the distorted shape which the theory of collisional transport predicts when there is heat flux.

28.2   Runaway Electrons and Dreicer Field

A classic example of transition from collisionality to collisionless behaviour is the phenomenon of runaway electrons, which appears when a plasma is in a strong electric field (aligned to the magnetic field if the plasma is magnetized). Let us model the runaway process simply by considering the motion of an electron in an electric field E, collisional effects being represented by a velocity dependent friction force, so that the dynamic equation for the electron can be written as:

$$m\frac{dv}{dt} = qE - m\,\nu(v)\,v \tag{205}$$

m and q being its mass and electric charge, and $\nu(v)$ the collisin frequency. When v is less than the thermal velocity, it is almost constant, whereas when v becomes larger it decreases as $1/v^3$. If any limit velociy can be reached at all, it must be given by the equation

$$v\,\nu(v) = \frac{q}{m}E \tag{206}$$

The function $v\,\nu(v)$ increases as v for subthermal velocities, and decreases as $1/v^2$ for suprathermal ones. It reaches a maximum for an almost thermal velocity. If qE/m less than this maximum value, there are two solutions, the smallest one being first reached and stable. The limit velocity of the electron then is

$$v \approx qE/(m\nu(v_T)) \tag{207}$$

and the electric current density is approximately:

$$j = nqv = \frac{nq^2}{m\nu(v_T)}\,E \tag{208}$$

One can read on this expression the value of the electrical conductivity. If on the other hand the electric field is so intense that qE/m exceeds the maximum value of the function $v\nu(v)$, there is no limit velocity, and the electron is indefinitely accelerated. This is because in a thermal collision time an electron receives so much acceleration that it exceeds the thermal speed and enters a regime in which it is less coupled by collision to the plasma. This allows acceleration for a time longer than the thermal collision time, therefore the electron picks up a supplement of kinetic energy, which makes it even less collisionally coupled, and so on. The corresponding electric field value is the runaway field, which has a value very close to the so called Dreicer field E_D. Precise definitions are given by Benz (1993). An estimate is obtained by equating qE/m and the maximum value of $v\,\nu(v)$ which is close to $v_T\,\nu(v_T)$:

$$E_D \approx \frac{mv_T}{q\nu(v_T)} \tag{209}$$

When the electric field approaches the Dreicer limit by lower values, the electrons in the wings of the thermal distribution already start suffering the runaway effect which affects most electrons when the runaway field is eventually reached.

28.3   Microinstabilities and Anomalous Transport Phenomena

When the distribution function of electrons or ions is shaped in a way which differs strongly from the thermal distribution or when the bulk velocity of one species with respect to another one exceeds some critical velocity, for example a thermal one, it often happens that the plasma turns micro-unstable. Micro-instabilities consist in the unstable growth in the Vlassov regime of electric or magnetic fluctuations in the plasma, usually of a frequency exceeding, sometimes by large factors, both the collision frequency and the ion gyrofrequencies. The growth is due to the fact that their Landau or cyclotron collisionless damping becomes negative. This introductory lecture is not the place to describe in detail how microinstabilities are triggered but it must be kept in mind that they may have most important effects on the global plasma dynamics. The phase of growth and non linear saturation, if any, is entirely described in the collisionless regime. Once developped, these fluctuations may be felt by the plasma particles as random scatterers, the time scale over which they affect particles velocities being much shorter than the collision time scale. From a macroscopic point of view everything appears as if in the region where these fluctuations are present the plasma had acquired a much higher degree of collisionality than the one allowed to it by normal Coulomb collisions in a quiet medium. This has a tremendous effect on transport properties. Microinstabilities may be triggered for example if the current density exceeds some threshold (see for example Heyvaerts 1981), the plasma then becoming anomalously resistive because the electrons suffer "collisions" on the fluctuations at a rate much larger than Coulomb collision frequency.

Else, the micro-instability may be triggered by the heat flux exceeding some threshold value, the mean free path of the electrons being suddenly reduced to a much smaller value by unstable fluctuations. This may reduce the field-aligned thermal conductivity, but may also increase it perpendicular to field lines, because the latter is limited by the fact that the electron's excursions perpendicular to the field is limited to no more than a gyroradius between two successive collisions. In the presence of microinstabilities, the "collisions" become a lot more "frequent", and this component of the heat conductibility may be increased as compared to the case of Coulomb collision transport.

28.4   Collisionless Shock Waves

Microinstabilities play a prominent role in the phenomenon known as "collisionless shock waves" (Sagdeev 1960, 1966, Galeev 1976). In the hydrodynamic regime a shock develops when for example a sound wave of finite and not too small amplitude propagates. It is a classical exercise to show that in one-dimensional propagation non linear effects drive the flow to progressive gradient steepening that culminate in the formation of a velocity and density discontinuity (see for example a pedagogical discussion of this in Zeldovitch and Raizer (1967) or Heyvaerts (1991)). This phenomenon can be viewed as an attempt by faster fluid in the trailing part of the wave to surpass the slower fluid moving in front of it. This is however impossible in hydrodynamic regime, because collisions severely forbid the propagation through one fluid component of another much faster one. A double-peaked distribution function could only possibly develop on a scale of one or a very few mean free paths, but is then thermalized away by collisions. The closer one can come to interpenetrating flows in the hydro regime is to be found in a quasi-discontinuity, i.e. in a region where the hydrodynamic quantities change extremely rapidly on a scale of a few mean free paths. This is an hydrodynamic, or collisional, shock wave. Hydrodynamics is of course unable to accurately describe what is going on in the few mean free paths that constitute the so called shock front. In hydrodynamic, or MHD, theory, this region is treated as a true discontinuity. This may look odd, but in fact a precise description of what really is going on in this very small region is not needed, because the preshock state and the postshock state of the fluid are related by very general conservation relations known as Rankine-Hugoniot relations (for a pedagogical presentation see Heyvaerts (1991) or Priest (1982)). In a tenuous plasma, Coulomb collisions are much less efficient and the Coulomb mean free path may be quite large. The invasion of a fast plasma stream into more slowly moving plasma is then not so inconceivable and the interpenetration region may not be so small. Regions where the distribution functions become double-peaked, each peak corresponding to one component of the plasma, the fast stream or the slow stream, or at least regions where the distribution function becomes very different from a thermodynamical equilibrium one, may well exist. Similar extreme situations may create very non-thermal electron distributions or strong differential motions between ions and electrons. Then one should question the microstability of such plasma states. When two plasma components have large streaming velocity one with respect to the other, the development of a micro-instability is likely, and its development may lead to a state of anomalous "collisionless" friction between the two components. Ultimately a state of recovered stability will be reached after the micro-turbulent episode, possibly with distribution functions exhibiting much broader dispersion than in the pre-turbulent state. The scale length on which this anomalous frictional braking and heating develops may be much less than the Coulomb mean free path, and is determined by the characteristic saturation scales of the microinstability. The thickness of this microturbulent region plays the role of a shock front thickness because it may be small when viewed on the system's global scale. Actually it is the thickness of the transition region between the upstream state, where the plasma streams had not yet interpenetrated, and the downstream state where they have been brought to a common velocity and to an higher temperature. Such a transition structure where the friction between some plasma components is mediated by micro-instabilities is called a collisionless shock wave. For example, the bow shock driven by earth in the flow of the solar wind is of a collisionless nature.

References

Balescu, R., 1975, "Equilibrium and non-equilibrium statististical mechanics", John Wiley.

Benz, A.O., 1993, Plasma Astrophysics, Kluwer Acad. Press.

Berger, M. and Field, G., 1984, Jounal Fluid Mech, **147**, 133.

Braginskii, 1965, Reviews of Plasma Physics, Vol **1**, Consultants Bureau, New York.

Cowling, T.G., 1957, Magnetohydrodynamics, Interscience, London.

Delcroix J.L., 1963, Physique des Plasmas I, Dunod, Paris.

Delcroix J.L.,1966 Physique des Plasmas II, Dunod, Paris.

Galeev, A., 1976, "Collisionless shocks", in Physics of Solar and Planetary Environnement, D.J. Williams ed., p. 464, AGU, Washington DC.

Huang, K., 1963, Statistical Mechanics, John Wiley.

Heyvaerts J., 1981, in "Solar Flare Magnetohydrodynamics", E.R. Priest ed., page 429-551, Gordon and Breach.

Heyvaerts J., 1991 in "Late stages of Stellar evolution", C. de Loore ed., EADN School held at Ponte de Lima, Lecture notes in physics **373** p 313, Springer Verlag.

Ichimaru, S., 1973, "Basic principles of plasma physics", Benjamin, Reading, Mass. USA.

Kalman, G., 1978, "Strongly Coupled Plasmas", NATO advanced study institutes series, Plenum Press, NY and London

Parnell, C., in these proceedings.

Priest E.R., 1982, "Solar Magnetohydrodynamics", Geophysics and Astrophysics Monographs, D. Reidel Publishing Comp.

Sagdeev, R.Z., 1960, "Shock waves in rarefied plasmas", Proc. 4th ICIPG, Vol 2, North Holland, Amsterdam.

Sagdeev, R.Z., 1966, in Reviews of Plasma Physics, 4 "Cooperative phenomena and shock waves in collisionless plasmas".

Spitzer, L., 1962, Physics of fully ionized gases, Interscience publishers, New York.

Uelenbeck, G.E. and Ford, G.W., 1961, in "Lectures in Statistical Mechanics" Lecture Notes in Applied Maths , Vol 1, American Math. Soc., Providence, Rhode Island.

Woltjer, L., 1958, Bull. Astron. Netherlands 14 , 39

Yvon,J., 1965, "Les corrélations et l'entropie en mécanique statistique", Dunod, Paris.

Zeldovitch,Y.B. and Raizer, Y.P., 1967, Physics of Shock Waves and High Temperature Hydrodynamic Phenomena, Academic Press.

Magnetic Reconnection: Classical Aspects

Clare E Parnell

St. Andrews University, Department of Mathematics, North Haugh, St Andrews, Fife, KY16 9SS, SCOTLAND

Abstract. Magnetic reconnection is an important mechanism in astrophysics for converting magnetic energy to both thermal energy and bulk acceleration of plasma and also for changing the global topology of the magnetic field. For over 50 years now solar theorists have investigated reconnection. This paper provides a basic review of the classical aspects of reconnection in both one, two and three dimensions, as well as, giving a potted history of reconnection theory in solar physics. Magnetic annihilation, Sweet-Parker reconnection and Petschek reconnection will all be discussed as will spine and fan reconnection in three dimensions.

1 Introduction

One of the most important mechanisms in astrophysics is magnetic reconnection because it allows global changes in the topology of the magnetic field. Such changes are, of course, associated with energy releases: reconnection is an efficient means of converting magnetic energy to thermal energy, bulk kinetic energy and accelerated particle. It can create large electric currents, shock waves and filamentation.

Magnetic reconnection plays a major role in many areas of astrophysics including:

- the interaction of the solar wind with the Earth's magnetosphere,
- in the action of the Sun's dynamo in generating the Sun's magnetic field.
- in heating solar and stellar coronae,
- in solar flares and coronal mass ejections.
- in the acceleration of stellar and solar winds,
- in accretion disks

Clearly magnetic reconnection occurs in many solar and stellar events and is therefore a key mechanism that, if fully understood, could help us understand many astrophysical phenomena.

2 History of Solar Magnetic Reconnection

Over 50 years ago now Giovanelli [1] suggested that solar flares occurred near magnetic neutral points. This observation led Cowling [2] to consider whether solar flares could be due to ohmic heating. To investigate this idea he calculated that a current sheet just a few metres thick would be needed to explain the massive and very rapid energy releases from solar flares. Later that year, however, Dungey [3] pointed out that a current sheet could be formed by an instability near a neutral point and that magnetic fieldlines of force could be broken and rejoined in the neighbourhood of a neutral point. This was the first time reconnection was eluded to, however, it was 5 years later before this process was called reconnection. In 1957, Sweet conceived an idea for a simple model of magnetic reconnection where the inflow into and outflow diffusion from a current sheet are balanced leading to a steady state with constant current and with the help of Parker the scaling laws of such a model were calculated [4,5]. Parker [6] went on to extend this model by including compressibility and internal structure across the sheet. It was in this paper that he coined the phrase 'annihilation of magnetic field' and discovered that the mechanism of reconnection proposed was too slow by a factor of 100 to explain the rapid energy releases from solar flares. This was a blow to the solar physics community who wanted to believed that reconnection was the answer to the heating of solar flares. Furthermore, Furth, Killeen and Rosenbluth [7] demonstrated that a current sheet was unstable to reconnection due to the tearing mode instability, implying that long Sweet-Parker diffusion regions were unlikely to exist. A year later Petschek [8] presented a mechanism for fast reconnection in which energy was released not only by ohmic heating, but also from pairs of shocks on both ends of the current sheet. Calm was again restored to the solar physics community as they once more believed they understood solar flares.

This feeling lasted for more than 20 years then Biskamp [9] published a paper describing numerical experiments in which he had attempted to reproduce Petschek reconnection and failed. He found no evidence of fast reconnection at high magnetic Reynolds numbers and therefore announced that fast reconnection could not occur in the solar corona. However, careful scrutiny of Biskamps experiment shows that it was not exactly the same as Petscheks model. Indeed, the boundary conditions imposed on his numerical box were different and so a different solution is found [10]. This was later demonstrated numerically by Priest and Lee [11]. Also Scholer [12] showed that at high magnetic Reynolds numbers fast Petchek reconnection can occur when the appropriate boundary conditions are imposed and the central current sheet possesses an anomalous resistivity.

Although work on 2D reconnection has continued since then both along analytical and numerical lines many have turned to three-dimensional reconnection. Schindler et al. [13] and Hesse and Schindler [14] were amongst the first and they attempted to address the problem of the definition of 3D reconnection, which is not the same as in 2D. This argument still continues today [15]. Lau and Finn [16], Priest and Demoulin [17] and Priest and Titov [18] amongst others, have considered various models for reconnection in 3D including ones for reconnection with and without neutral points.

3   Review of MHD Equations

Before the classical models of reconnection are explained lets take a brief look at the MHD equations so we understand a little better about where and how reconnection can take place.

In the equations of magnetohydrodynamics (MHD) it is assumed that the typical length scales are much greater than the particle collision mean free path and that the typical velocities involved are much less than the speed of light (i.e. nothing is moving relativistically). This means that Maxwells equations and Ohm's law can be written in the form

$$\nabla \times \mathbf{B} = \mu_0 \mathbf{j} \, , \tag{1}$$

$$\nabla \times \mathbf{E} = -\frac{\partial \mathbf{B}}{\partial t} \, , \tag{2}$$

$$\mathbf{E} + \mathbf{v} \times \mathbf{B} = \frac{\mathbf{j}}{\sigma} \, , \tag{3}$$

$$\nabla \cdot \mathbf{B} = 0 \, , \tag{4}$$

where the magnetic permeability, $\mu_0 \approx 1.26 \times 10^{-6}$ Hm^{-1} and σ is the electrical conductivity. The first two equations are Ampere's law and Faraday's law, respectively, and the third equation is Ohm's law.

Induction Equation

If the curl of Ohm's law is taken then

$$\nabla \times \mathbf{E} + \nabla \times (\mathbf{v} \times \mathbf{B}) = \nabla \times \mathbf{j}/\sigma \, .$$

Substituting in for $\mathbf{E}$ and $\mathbf{j}$ from (2) and (1) we find

$$-\frac{\partial \mathbf{B}}{\partial t} + \nabla \times (\mathbf{v} \times \mathbf{B}) = \frac{1}{\sigma \mu_0}(\nabla \times (\nabla \times \mathbf{B})) = \eta(\nabla(\nabla \cdot \mathbf{B}) - \nabla^2 \mathbf{B}) \, ,$$

where $1/\sigma\mu_0 = \eta$ is the magnetic diffusivity.
When rearranged this gives the *Induction Equation*,

$$\frac{\partial \mathbf{B}}{\partial t} = \nabla \times (\mathbf{v} \times \mathbf{B}) + \eta \nabla^2 \mathbf{B} \, , \tag{5}$$

which describes the evolution of the magnetic field.

Let us consider the ratio of the terms on the RHS of (5),

$$\frac{\nabla \times (\mathbf{v} \times \mathbf{B})}{\eta \nabla^2 \mathbf{B}} = \frac{vB/l}{\eta B/l^2} = \frac{vl}{\eta} = R_m \, ,$$

where l and v are typical length and velocity scales, respectively. This ratio is equal to the *magnetic Reynolds number*, R_m.

In the solar corona, l is typically $10^6 - 5 \times 10^7$ m and v is typically $10^2 - 10^5$ ms^{-1} so with a magnetic diffusivity η of 1 m^{2}s^{-1} the magnetic Reynolds number becomes $10^8 - 5 \times 10^{12}$. This means that typically in the solar corona the magnetic Reynolds number is very large. What does this imply about the magnetic field?

$R_m \gg 1$

The limit $R_m \gg 1$ is known as the perfectly conducting limit: this does not mean that there is no current rather that the second term on the RHS of the induction equation is negligible. Thus, in such a limit the induction equation reduces to

$$\frac{\partial \mathbf{B}}{\partial t} = \nabla \times (\mathbf{v} \times \mathbf{B}) .$$

An important consequence of this is Alfven's Theorem.

Alfvens Theorem

Consider any surface S bounded by a closed contour C moving with the plasma at a local velocity $\mathbf{v}$ with a flux F crossing the surface.

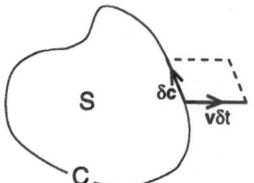

Fig. 1. A moving plasma element of surface S with a boundary C.

In time δt, the line element δc sweeps out an area $\mathbf{v}\delta t \times \delta \mathbf{c}$, so

$$\frac{DF}{Dt} = \partial F/\partial t + \mathbf{v} \cdot \nabla F = \int_S \frac{\partial \mathbf{B}}{\partial t} \cdot d\mathbf{S} + \oint_C \mathbf{B} \cdot \mathbf{v} \times d\mathbf{c} . \tag{6}$$

The rate of change of flux is equal to the changes in the magnetic field with time, the 1st term on the RHS of (6), and due to changes caused by the motion of the boundary, the 2nd term on the RHS of (6). This last term can be rewritten using Green's theorem as a surface integral,

$$\oint_C \mathbf{B} \cdot \mathbf{v} \times d\mathbf{c} = - \oint_C \mathbf{v} \times \mathbf{B} \cdot d\mathbf{c} = - \int_S \nabla \times (\mathbf{v} \times \mathbf{B}) \cdot d\mathbf{S} ;$$

thus the equation for the rate of change of flux becomes,

$$\frac{DF}{Dt} = \int_S \left(\frac{\partial \mathbf{B}}{\partial t} - \nabla \times (\mathbf{v} \times \mathbf{B}) \right) \cdot d\mathbf{S} = 0 ,$$

using the limit of the induction equation for high magnetic Reynolds number. This implies that flux (F) through surface (S) moving with the plasma is constant. And since this holds for any arbitrary contour C it implies that the magnetic field must move with the plasma, thus magnetic fieldlines are said to be *frozen-in* to the plasma (Figure 2).

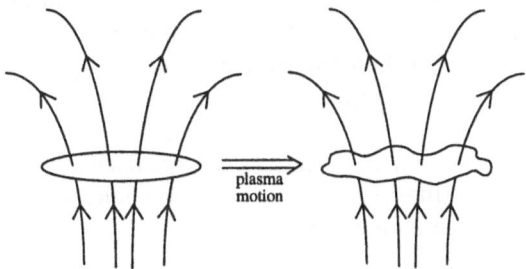

Fig. 2. Magnetic flux conservation - if a curve C_1 is distorted into C_2 by plasma motion, then the flux through C_1 equals the flux through C_2

That is to say plasma elements can only move along fieldlines and can not cross from one fieldline to another (Figure 3).

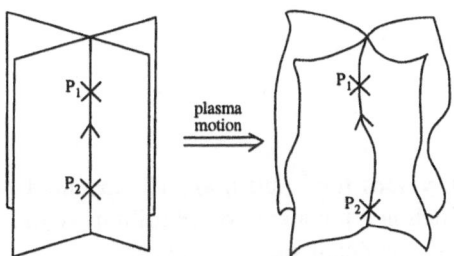

Fig. 3. Magnetic fieldline conservation - if plasma elements P_1 and P_2 lie on one fieldline at time t_1 then they will lie on the same fieldline at time t_2

This approximation is valid for most solar MHD situations. In the corona magnetic forces dominate so the plasma is pulled along by the magnetic field. In the photosphere the inertia of plasma dominates and magnetic field is dragged by plasma.

The transport of the magnetic field by the fluid is known as magnetic advection and so the term $\nabla \times \mathbf{v} \times \mathbf{B}$ in induction equation is known as the *advection term*.

$R_m \ll 1$

A magnetic Reynolds number much less than 1 implies that length scales are small. In such a situation the induction equation reduces to

$$\frac{\partial \mathbf{B}}{\partial t} = \eta \nabla^2 \mathbf{B} .$$

This is in the form of a diffusion equation and represents the ability of the magnetic field to move through the plasma (diffuse). Hence η is known as the diffusion coefficient. When the diffusion term is dominant reconnection may take place; thus this implies that small length scales are needed for reconnection.

Small length scales can be found in many situations including: current sheets; neutral points; due to tearing mode and coalescence instabilities; and in propagating sheets in shock waves.

Diffusion Timescale

One of the key questions of interest is what is the timescale for magnetic diffusion?
From the diffusion equation we find that

$$\frac{B}{\tau} = \frac{\eta B}{l^2} ,$$

thus the diffusion timescale is,

$$\tau_{diff} = \frac{l^2}{\eta} .$$

Now typical coronal values for l and η are 10^7 m and 1 m^{2}s^{-1}, respectively, so $\tau_{diff} \approx 10^{14}$ s which is equivalent to 3.8 million years! Clearly diffusion in the solar corona does not occur on a global scale.

As we have seen for rapid diffusion we need very short length scales which implies steep magnetic field gradients and thus electric currents. Clearly current sheets are an ideal place for magnetic diffusion and therefore reconnection to occur.
Rearranging the above equation we find that the rate of diffusion is

$$v_{diff} = \frac{\eta}{l} \tag{7}$$

Globally in the solar corona v_{diff} is $\approx 10^{-7}$ ms^{-1} so is very slow.

4   One-Dimensional Magnetic Reconnection

The simplest type of reconnection that can be imagined is a one-dimensional process where a parallel magnetic field diffuses through a static plasma. This is known as magnetic annihilation. In such a situation we have

$$\mathbf{v} = (0,0,0) \,, \qquad \text{and} \qquad \mathbf{B} = (0, B(x,t), 0) \,.$$

This means the induction equation becomes

$$\frac{\partial B}{\partial t} = \eta \frac{\partial^2 B}{\partial x^2} \,.$$

If this equation is solved under the initial conditions

$$B(x,t) = \begin{cases} B_0 & x > 0 \\ -B_0 & x < 0 \end{cases},$$

it yields the following solution

$$B(x,t) = B_0 \mathrm{erf}\left(\frac{x}{\sqrt{4\eta t}}\right) = \frac{2B_0}{\sqrt{\pi}} \int_0^{x/\sqrt{4\eta t}} e^{\lambda^2} \mathrm{d}\lambda \,, \qquad (x > 0) \,,$$

and $B(-x,t) = -B(x,t)$.

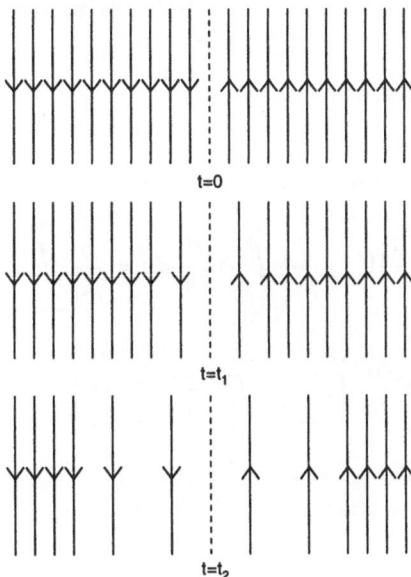

Fig. 4. The magnetic field $\mathbf{B}$ in the xy plane at time $t = 0, t_1$ and t_2

As the magnetic field diffuses it varies as shown in Figure 4 for times $t = 0$, t_1, t_2. Clearly the strength of the magnetic field near the x-axis decreases over time (Figure 5). This is because the oppositely directed fieldlines diffuse through the plasma and cancel (annihilate). Initially the magnetic field gradient at $x = 0$ is steep, but as the process of annihilation continues this gradient decreases, smoothes out as the field is diffused away, thus the current sheet widens and the diffusion rate, $v_{diff} = \eta/l$ decreases, since l the scale width of the current sheet increases.

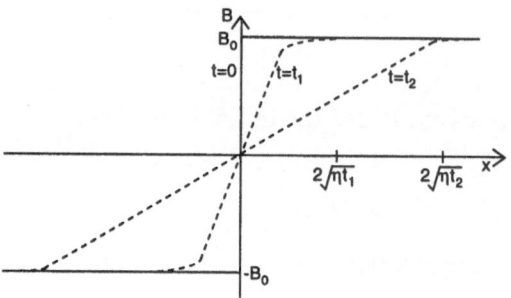

Fig. 5. The strength of the magnetic field B against x at time $t = 0, t_1$ and t_2

Magnetic Energy

What are the implications for the magnetic energy released from such a process? The magnetic energy is defined as

$$E_m = \frac{1}{2\mu_0} \int_V B^2 dV \ .$$

So the rate of change of magnetic energy is

$$\frac{\partial E_m}{\partial t} = \frac{\partial}{\partial t} \int_V \frac{B^2}{2\mu_0} dV = \frac{1}{\mu_0} \int_{-\infty}^{\infty} B \frac{\partial B}{\partial t} dx = \frac{\eta}{\mu_0} \int_{-\infty}^{\infty} B \frac{\partial^2 B}{\partial x^2} dx \ ,$$

$$= \frac{1}{\sigma \mu_0^2} \left\{ \left[B \frac{\partial B}{\partial x} \right]_{-\infty}^{\infty} - \int_{-\infty}^{\infty} \left(\frac{\partial B}{\partial x} \right)^2 dx \right\} \ .$$

At infinity $\partial B/\partial x$ remains equal to zero so the first term vanishes and from Ampere's law we know that $\mu_0 j = \partial B/\partial x$, thus

$$\frac{\partial E_m}{\partial t} = -\frac{1}{\sigma} \int_{-\infty}^{\infty} j^2 dx \ .$$

That is to say the magnetic energy is dissipated entirely through ohmic heating.

Clearly, in our magnetic annihilation model as the current sheet widens ohmic heating decreases with time so this type of model is no good as a flare model since it cannot explain the rapid heating observed in solar flares over long periods. We need both advection and diffusion for reconnection to go faster. Let us now move on to the classical reconnection models in 2D.

5 2D Magnetic Reconnection

5.1 Introduction

Two-dimensional reconnection is the process by which flux is transferred from one topologically distinct magnetic region to another. Alternatively, it can be defined as the mechanism by which new fieldlines are created.

Let us consider a 2D magnetic X-point. A point where $\mathbf{B} = \mathbf{0}$, near which the field is divided into four topologically distinct regions by special fieldlines known as separatrices. The separatrices extend from either a magnetic source or sink into the null. Consider the two fieldlines AB and CD indicated by a dashed and dotted line respectively in Figure 6. A flow across the separatrices brings the fieldlines in towards the null such that they lie along the separatrices. Then diffusion may occurs near the null where short length scales exist breaking and recombining the original fieldlines to give new fieldlines AD made up of the first part of AB and the last part of CD and CB made up of the first part of CD and the last part of AB.

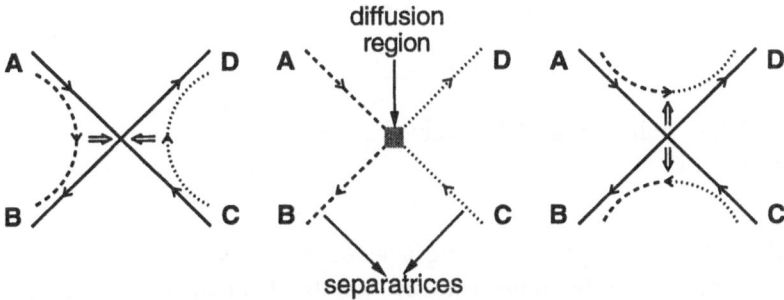

Fig. 6. Two-dimensional reconnection at an X-point of fieldlines AB (dashed) and CD (dotted) to form fieldlines AD and CB

In 2D, reconnection can only occur at null points. It occurs due to non-ideal effects and causes changes in the topology of the magnetic field and gives rise to plasma flows across separatrices.

5.2   2D Neutral Points

Structure

A magnetic neutral (null) point is a point at which all components of the magnetic field are equal to zero, $\mathbf{B} = \mathbf{0}$. The local field near such a point can be described by

$$\mathbf{B} = M \cdot \mathbf{r} \,,$$

assuming, without loss of generality, that the neutral point is at the origin, and where M is a 2x2 matrix and the position vector, $\mathbf{r} = (x, y)^T$. The simplest form for M [19] is

$$M = \begin{bmatrix} 0 & (q - j_z)/2 \\ (q + j_z)/2 & 0 \end{bmatrix} \,,$$

where q is some arbitrary constant and j_z is the current in the z direction,

$$\mathbf{j} = (0, 0, j_z) \,.$$

Since $\nabla \cdot \mathbf{B} = 0$ the sum of the eigenvalues of M equal zero, thus if λ_1 and λ_2 are the eigenvalues then

$$\lambda_2 = -\lambda_1 \,.$$

Assuming the eigenvalues are real and distinct the equation for a magnetic fieldline can be written,

$$\mathbf{r}(k) = Ae^{\lambda_1 k}\mathbf{x}_1 + Be^{\lambda_2 k}\mathbf{x}_2 \,,$$

where $\mathbf{x}_i$ is the eigenvector of M associated with eigenvalue λ_i.
   If $\lambda_1 = -\lambda_2 = \lambda > 0$ then as $k \to \infty$

$$\mathbf{r}(k) \to Ae^{\lambda k}\mathbf{x}_1 \,,$$

and fieldlines will lie parallel to the eigenvector $\mathbf{x}_1$.
However, as $k \to -\infty$

$$\mathbf{r}(k) \to Be^{-\lambda k}\mathbf{x}_2 \,,$$

and the fieldlines lie parallel to eigenvector $\mathbf{x}_2$.
   Furthermore the fieldlines that lie exactly along the eigenvectors $\mathbf{x}_1$ and $\mathbf{x}_2$ are the special fieldlines known as the separatrices. Real and distinct eigenvalues can arise if $|j_z| < |q|$ (Figure 7a and 7b). If $|j_z|=0$ then the separatrices are at right angles and the field is potential, whereas if $|q| > j_z| > 0$ then the separatrices are at less than 90 degrees. If the eigenvalues of the matrix M are repeated ($|j_z| = |q|$) a degenerate 1D field arises containing a null line (Figure 7c). Complex eigenvalues ($|j_z| > |q|$) give rise to elliptical fieldlines (Figure 7d).

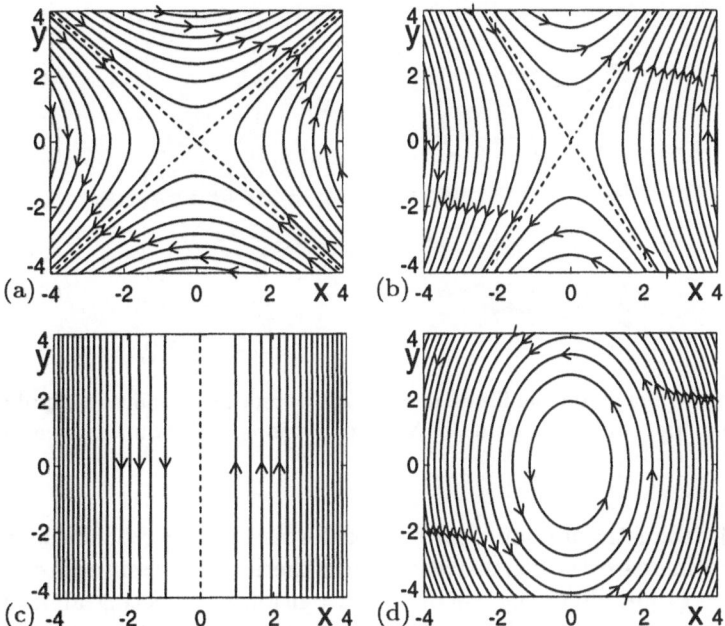

Fig. 7. Two-dimensional magnetic neutral points: (a) a potential null ($j_z = 0$), (b) non-potential X-point ($|j_z| < q$), (c) anti-parallel fieldlines ($|j_z| = q$) and (d) elliptic null ($|j_z| > q$)

Collapse

Dungey [3] was the first to investigate the collapse of an X-point. He considered the potential magnetic field $\mathbf{B} = (y, x)$ which has fieldlines

$$y^2 - x^2 = \text{constant}.$$

The separatrices are the fieldlines that pass through the null which is at the origin, so they are given by $y = x$ and $y = -x$. Since the null is potential, $\mathbf{j} = 0$ and the magnetic pressure and tension forces balance, $\mathbf{j} \times \mathbf{B} = 0$ (Figure 8a).

If we now perturb the null slightly so that

$$\mathbf{B} = (y, (1 + \alpha^2)x) \qquad \alpha^2 \ll 1\,,$$

the fieldlines become

$$y^2 - \alpha^2 x^2 = \text{constant}\,.$$

The null is still at the origin, but now the separatrices are given by $y = \alpha x$ and $y = -\alpha x$ and the field is now no longer potential, but has current,

$$\mathbf{j} = (0, 0, (\alpha^2)/\mu_0)\,.$$

This means that the Lorentz force is non-zero,

$$\mathbf{j} \times \mathbf{B} = (-(1 + \alpha^2)\alpha^2 x/\mu_0, \alpha^2 y/\mu_0, 0)\,,$$

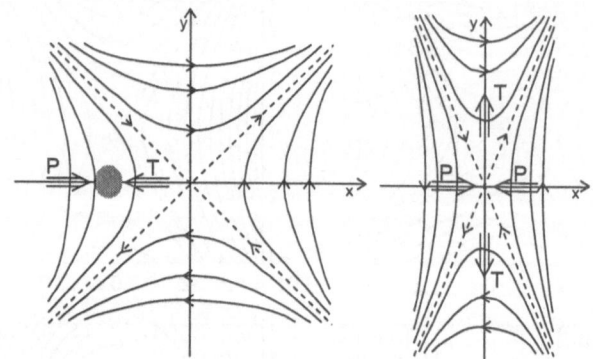

Fig. 8. (a)The magnetic fieldlines near a 2D potential X-point in equilibrium. (b)The magnetic fieldlines near a 2D X-point not in equilibrium due to a uniform current

and so the magnetic pressure and tension forces no-longer balance. In fact, they act to increase the perturbation since there is a pressure force acting towards the origin along the x-axis and a tension force acting along the y-axis away from the origin (Figure 8b). Hence the X-point is unstable. Note, however, this is only true since the fieldlines are assumed to be free to move and are not line-tied at the boundaries, thus energy can propagate into the system.

Formation of a Current Sheet

It was following this principle that Green [20] and Syrovatsky [21] used a kinematic approach to explain the formation of a current sheet at a neutral point. In a kinematic approach it is assumed that

$$\mathbf{E} + \mathbf{v} \times \mathbf{B} = \mathbf{0} \,, \tag{8}$$

$$\nabla \times \mathbf{E} = \mathbf{0} \,. \tag{9}$$

Let us assume that our initial magnetic field is the potential field $\mathbf{B} = (y, x)$, thus $\mathbf{j} \times \mathbf{B} = \mathbf{0}$. The plasma flow is assumed to be of the form

$$\mathbf{v} = (v_x(x, y), v_y(x, y)) \,,$$

and from the kinematic equations we therefore find that

$$\mathbf{E} = (0, 0, E_0).$$

Now taking (8)$\times\mathbf{B}$ we find that the component of velocity perpendicular to the field B is equal to

$$\mathbf{v}_\perp = \frac{\mathbf{E} \times \mathbf{B}}{B^2} = \left(-\frac{E_0 x}{x^2 + y^2}, \frac{E_0 y}{x^2 + y^2} \right) \,.$$

The magnitude of $\mathbf{v}_\perp$ is

$$|\mathbf{v}_\perp| = E_0/B = E_0/(x^2 + y^2)\,,$$

and so it is singular at the origin. Since our flow can not be infinite one of our assumptions must have broken down. Ohm's Law (8) breaks down at the origin, hence $\mathbf{j} = 0$ everywhere, save at origin, where a current sheet may form (Figure 9).

The easiest way to represent a current sheet is as a cut in the complex plane. So if we write $z = x + iy$ then initially,

$$B(z) = B_y + iB_x = z.$$

and the current sheet field becomes

$$B(z) = B_y + iB_x = (z^2 + l^2)^{\frac{1}{2}}.$$

This represents a current sheet that extends from $-l < y < l$.

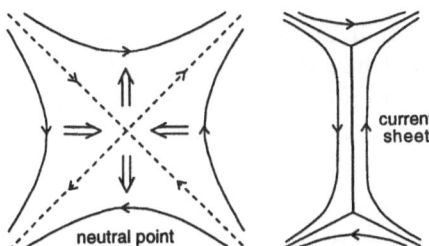

Fig. 9. (a)The magnetic field about a potential 2D X-point. (b)The magnetic field produced by slow motions indicated in (a) by the double arrows under the perfectly conducting assumption

5.3   Current Sheets

A current sheet is a non-propagating boundary between two plasmas. On either side the magnetic field is transverse to boundary, hence they are a form of tangential discontinuity. There is no velocity flow across the boundary which, since it is stationary, must be in total pressure balance, hence

$$p_1 + \frac{B_1^2}{2\mu_0} = p_2 + \frac{B_2^2}{2\mu_0}\,,$$

where p_1 and $\mathbf{B}_1$ are the plasma pressure and magnetic field on one side of the current sheet, and p_2 and $\mathbf{B}_2$ are the plasma pressure and magnetic field on the other side.

A neutral sheet is one in which $\mathbf{B} = 0$, however, not all current sheets are neutral (see Figure 10). In the interior of a current sheet the magnetic field evolves through diffusion whereas, outside the sheet, the magnetic field no longer diffuses, but advects and so it's evolution is defined by ideal MHD.

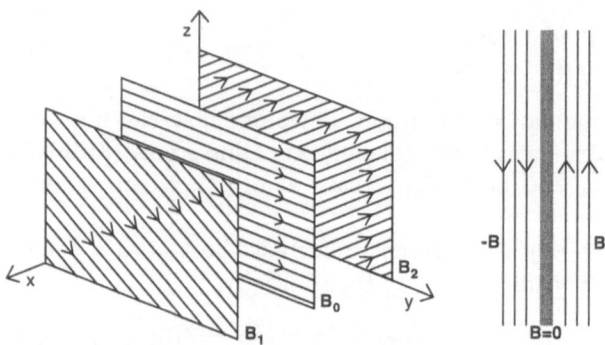

Fig. 10. (a)The magnetic field in planes either side of a current sheet with a uniform field B_0 in. (b)The magnetic field either side of a neutral sheet

5.4   The Rate of Reconnection

The key point of interest in any reconnection model is the rate at which the reconnection occurs, but how do we calculate it?

The rate of reconnection is the rate of change of flux which equals $\partial A/\partial t$, where A is the flux function which satisfies

$$\mathbf{B} = \nabla \times \mathbf{A} .$$

Hence from Faradays law (2) we find that

$$\nabla \times \mathbf{E} = -\frac{\partial \mathbf{B}}{\partial t} = -\frac{\partial(\nabla \times \mathbf{A})}{\partial t} ,$$

which implies

$$\mid \mathbf{E} \mid = \mid \frac{\partial \mathbf{A}}{\partial t} \mid ,$$

and so the magnitude $\mid \mathbf{E} \mid$ is proportional to the rate of reconnection. If the reconnection is steady $\partial \mathbf{B}/\partial t = 0$, and $\mathbf{E} = $ constant, therefore the rate of reconnection is constant. From Ohm's law,

$$\mid \mathbf{E} \mid = \mid \mathbf{v} \times \mathbf{B} \mid = vB = M_A B v_A ,$$

where $M_A = v/v_A$ is the Alfven Mach number. If the magnetic field is uniform then Bv_A is constant and therefore the inflow Mach number M_{Ai} may be used as a dimensionless measure of the reconnection rate.

If the reconnection rate is less than or equal to $R_m^{-1/2}$ as R_m tends to infinity, then it is classed as very slow or slow reconnection, respectively. However, if the reconnection is greater than $R_m^{-1/2}$ in the limit $R_m \to \infty$ it is classed as fast reconnection.

5.5   Steady-State Slow Reconnection: Sweet-Parker

The first type of 2D reconnection we will consider is the classic steady-state slow reconnection mechanism known as Sweet-Parker reconnection. Sweet [4] and Parker [5] described an order of magnitude theory for a diffusion region such as the one sketched in Figure 11. They assumed that the current

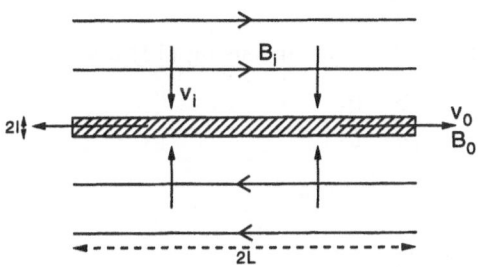

Fig. 11. Sketch of a Sweet-Parker diffusion region and surrounding field

sheet(diffusion region) is a thin layer of length $2L$ and width $2l$ ($l \ll L$) between two oppositely directed uniform magnetic fields. The plasma flow into the current sheet is v_i and ϱ_i is the inflow density. The outflow velocity and density are given by v_o and ϱ_o, respectively. The inflow magnetic field is assumed uniform and of magnitude B_i and outflow magnetic field of magnitude B_0. They assumed that the reconnection process was incompressible, thus $\varrho_i = \varrho_o = \varrho$ and that it was steady, so from the induction equation we find that the inflow velocity must be equal to the diffusion rate, hence

$$v_i = \frac{\eta}{l} \,. \tag{10}$$

Conservation of mass implies that the amount of plasma entering the diffusion region must equal the amount plasma leaving, thus

$$\varrho_i v_i / l = \varrho_o v_o / L \qquad \Rightarrow \qquad L v_i = l v_o \,. \tag{11}$$

Eliminating l from (10) and (11) gives

$$v_i = \sqrt{\frac{\eta v_o}{L}} \,. \tag{12}$$

Furthermore, the steady state Ohm's Law implies that the flux entering the sheet must equal the flux leaving the sheet, hence,

$$v_i B_i = v_o B_o \,. \tag{13}$$

From Ampere's Law (1) we find that the current in the sheet is

$$j_z = \frac{B_i}{l \mu_0} \,, \tag{14}$$

and $\nabla \cdot \mathbf{B} = 0$ implies

$$\frac{B_i}{L} = \frac{B_o}{l} . \tag{15}$$

The equation of motion along the sheet is just in the x-direction and so

$$\varrho(\mathbf{v}.\nabla)v_x = -\frac{\partial p}{\partial x} + j_z B_y .$$

Neglecting the pressure gradient and using (14) and (15) it reduces to

$$\varrho\left(\frac{v_o}{L}\right)v_o = \frac{B_i}{l\mu_0}\frac{lB_i}{L} ,$$

hence,

$$v_o^2 = \frac{B_i^2}{\mu_0\varrho} = v_{Ai}^2 , \tag{16}$$

where v_{Ai} is the inflow alfven speed.

Thus we now know outflow speed and how the inflow speed depends on v_o. So we can calculate the reconnection rate, $M_{Ai} = v_i/v_{Ai}$. From (12) $v_i = \sqrt{\eta v_o/L}$, (16) $v_0 = v_{Ai}$ and $\mathrm{R}_{mi} = Lv_{Ai}/\eta$ the Mach number becomes

$$M_{Ai} = \frac{v_i}{v_{Ai}} = \sqrt{\frac{\eta v_{Ai}}{L}}/v_{Ai} = \sqrt{\frac{\eta}{Lv_{Ai}}} = \frac{1}{\sqrt{\mathrm{R}_{mi}}} .$$

So the Sweet-Parker reconnection rate is $M_{Ai} = 1/\sqrt{\mathrm{R}_{mi}}$. If L is equal to a global length scale of 10^7 m, say, then $\mathrm{R}_{mi} \approx 10^{12}$ and so $M_{Ai} \approx 10^{-6}$ which is far too slow to explain solar flares!

5.6   Steady-State Fast Reconnection: Petschek

The other best known classical 2D reconnection is Petschek reconnection [8]. This is a fast steady-state reconnection mechanism that considers both the immediate field around diffusion region, plus the external field in the locality of the diffusion region. Petschek realised that slow magnetoacoustic shock waves provide another way, in addition to a diffusion region, of converting magnetic energy into heat and kinetic energy, but how could these shocks be formed? Slow magnetoacoustic shocks are generated when there is an obstacle in a flow which is traveling faster than the slow magnetoacoustic wave speed. In a reconnection situation the obstacle is the tiny diffusion region.

Petschek's analysis was not completely rigorous, but it did show great physical insight. It is quite hard to follow since it includes diffusion theory, potential theory and MHD shock theory, so here I do not present Petscheks analysis exactly as he did, but instead use an approach along similar lines.

Firstly, as shown in Figure 12, we denote the external magnetic field, flow and density by B_e, v_e and ϱ_e, respectively. The same quantities in the inflow

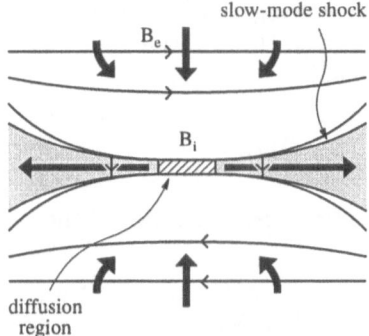

slow-mode shock

B_e

B_i

diffusion
region

Fig. 12. A sketch of Petschek's magnetic field configuration

region are denoted by B_i, v_i and ϱ_i and in the outflow region as B_o, v_o and ϱ_o, respectively. The small diffusion region is assumed to be a Sweet-Parker type region with width l and length L where, $l \ll L$ as before. The external region has a typical length scale of L_e where $L \ll L_e$. The external magnetic field is assumed to be uniform so $\mathbf{B_e} = (B_e, 0, 0)$.

The basic aim of the following analysis will be to calculate the maximum reconnection rate which implies finding the maximum M_{Ae} and hence the maximum external flow, v_e.

Relations Between External and Inflow Field

If we first consider the relations between the external and inflow field we find that since the system is assumed to be steady state Ohm's law implies

$$v_i B_i = v_e B_e ,$$

however,

$$v = M_A v_A = M_A \frac{B}{\sqrt{\mu_0 \varrho}} ,$$

so the above relation can be written

$$M_{Ai} B_i^2 = M_{Ae} B_e^2 . \tag{17}$$

Relations Across the Diffusion Region

Relations between the inflow and outflow parameters are simply the same as those in the Sweet-Parker model since the diffusion region is assumed to be a Sweet-Parker region. Hence from the steady-state assumption (10) $v_i = \eta/l$, and from conservation of mass (11) and (16) imply

$$L v_i = l v_o = l v_{Ai} .$$

If we now introduce the following dimensionless parameters: the external magnetic Reynolds number, R_{me}; the external and inflow alfven Mach numbers, M_{Ae} and M_{Ai}, which equal,

$$R_{me} = \frac{L_e v_{Ae}}{\eta}, \qquad M_{Ae} = \frac{v_e}{v_{Ae}} \qquad \text{and} \qquad M_{Ai} = \frac{v_i}{v_{Ai}},$$

then we can write the ratio of length scales of the system as

$$\frac{l}{L_e} = \frac{\eta}{v_i L_e} = \frac{v_{Ae}}{R_{me} v_i} = \frac{v_{Ae} B_i}{R_{me} v_e B_e} = \frac{M_{Ae}^{\frac{1}{2}}}{R_{me} M_{Ae} M_{Ai}^{\frac{1}{2}}} = \frac{1}{R_{me} M_{Ae}^{\frac{1}{2}} M_{Ai}^{\frac{1}{2}}}, (18)$$

$$\frac{L}{L_e} = \frac{l}{L_e} \frac{v_{Ai}}{v_i} = \frac{1}{R_{me} M_{Ae}^{\frac{1}{2}} M_{Ai}^{\frac{3}{2}}}. \tag{19}$$

Relations Across the Shock

We now consider the shocks that extend out from each corner of the diffusion region approximately parallel to the x axis. Due to the steady-state assumption they must remain stationary and so the shocks must be traveling at the same speed, but in the opposite direction to, the medium they are sitting in, hence

$$v_s = v_e .$$

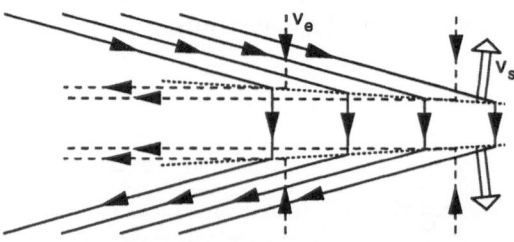

Fig. 13. Two slow mode MHD shock waves (dotted) back to back can make a field (solid) reversal region. The shock and field are traveling at speeds v_s and v_e, respectively.

The shock is said to be 'switch-off' in nature because the magnetic field behind shock is normal to the shock itself, therefore, from the 'switch-off' shock relations

$$v_e = v_s = \frac{B_n}{\sqrt{\mu_0 \varrho}}, \tag{20}$$

where B_n is the component of the magnetic field ahead of shock.

The effect of the shock is to slightly distort the external magnetic field in the inflow region so $B_i = B_e + B_{distort}$.

Since the inflow field is potential to first order the distortion to the external field can be regarded as begin produced by a series of monopoles along the x-axis between $-L_e$ and $-L$ and L and L_e (Figure 14). To lowest order the inclination of the shocks is neglected and at large distances the effect of the monopoles is neglected since the field from the monopoles drops off as $1/r$. The distortion of the inflow field is therefore equal to the sum of the

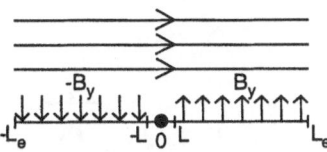

Fig. 14. The effective magnetic field of Petscheks reconnection site with the field due to the shocks modeled as monopoles along the x-axis between $(-L_e, -L)$ and (L, L_e).

effects of all the monopoles of strength B_y,

$$
B_{distort} = \int_{-L_e}^{-L} \frac{-B_y(x-a)}{\pi((x-a)^2 + y^2)} + \frac{-B_y y}{\pi((x-a)^2 + y^2)} da
$$

$$
+ \int_{L}^{L_e} \frac{B_y(x-a)}{\pi((x-a)^2 + y^2)} + \frac{B_y y}{\pi((x-a)^2 + y^2)} da \, ,
$$

$$
= \frac{B_y}{\pi} \int_{x+L_e}^{x+L} \frac{w}{w^2 + y^2} dw + \frac{B_y}{\pi} \int_{x+L_e}^{x+L} \frac{y}{w^2 + y^2} dw
$$

$$
- \frac{B_y}{\pi} \int_{x-L}^{x-L_e} \frac{w}{w^2 + y^2} dw - \frac{B_y}{\pi} \int_{x-L}^{x-L_e} \frac{y}{w^2 + y^2} dw \, ,
$$

$$
= \frac{B_y}{2\pi} \left[\log\left((x+L)^2 + y^2\right) - \log\left((x+L_e)^2 + y^2\right) \right.
$$

$$
\left. - \log\left((x-L_e)^2 + y^2\right) + \log\left((x-L)^2 + y^2\right) \right]
$$

$$
+ \frac{B_y}{\pi} \left[\tan^{-1}\left(\frac{x+L}{y}\right) - \tan^{-1}\left(\frac{x+L_e}{y}\right) \right.
$$

$$
\left. - \tan^{-1}\left(\frac{x-L_e}{y}\right) + \tan^{-1}\left(\frac{x-L}{y}\right) \right] \, .
$$

We are only interested in $B_{distort}$ in the immediate region about the diffusion regions (near the origin) where

$$B_{distort}(0,0) = \frac{2B_y}{\pi} \log \left(\frac{L}{L_e} \right) .$$

Now we need to determine B_y. Figure 15 shows a close up of one quadrant of the diffusion region and shock. Clearly for small θ,

$$B_n = B_y \cos \theta - B_e \sin \theta = B_y - B_e \theta .$$

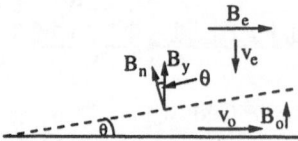

Fig. 15. A close up of one quadrant of a Petschek reconnection configuration with shock front indicated by a dashed line.

The conservation of B_n across shock implies

$$B_n = B_o \cos \theta \approx B_o ,$$

and the conservation of v normal across shock implies

$$v_e \cos \theta = v_o \sin \theta , \qquad \Rightarrow \qquad v_e = v_o \theta .$$

These two relations couple with Ohm's Law to give

$$v_e B_e = v_o B_o = v_o B_n = \frac{v_e B_n}{\theta} .$$

So $B_y = B_n + B_e \theta = 2B_n$.
Thus, the inflow magnetic field is equal to

$$B_i = B_e + \frac{4B_n}{\pi} \log \frac{L}{L_e} = B_e \left(1 + \frac{4M_{Ae}}{\pi} \log \frac{L}{L_e} \right) , \tag{21}$$

since from (20) $M_{Ae} = B_n/B_e = \theta \ll 1$.

Also, we know that $B_i/B_e \approx 1$ since the inflow field is almost uniform and therefore (18) and (19), the length scale relations, become

$$\frac{l}{L_e} = \frac{1}{R_{me} M_{Ae}} ,$$

$$\frac{L}{L_e} = \frac{1}{R_{me} M_{Ae}^2} .$$

This implies that the dimensions of the central current sheet decrease as the magnetic Reynolds number or reconnection rate M_{Ae} increases.

Petschek suggested that the mechanism chokes itself off when B_i becomes too small so he estimated a maximum reconnection rate at $B_i = B_e/2$

$$\max M_{Ae} = \frac{\pi}{8 \log \mathrm{R_{me}}} .$$

This would imply a typical maximum reconnection rate of $M_{Ae} \approx 0.01$. A rate that is fast enough to explain solar flares.

5.7   Steady-State Reconnection: Almost-Uniform

So far we have looked at the two classical steady-state reconnection mechanisms, but are there any more steady-state mechanisms? Yes, in fact there is a whole family of almost-uniform steady-state reconnection regimes [10]. These can be found by varying the boundary conditions of a setup similar to that of Petscheks.

Priest and Forbes [10] started with the equations for 2D incompressible flow,

$$\varrho(\mathbf{v} \cdot \nabla)\mathbf{v} = -\nabla p + (\nabla \times \mathbf{B}) \times \mathbf{B}/\mu_0 \tag{22}$$

$$\mathbf{E} + \mathbf{v} \times \mathbf{B} = 0 \tag{23}$$

$$\nabla \times \mathbf{E} = 0 \tag{24}$$

with $\nabla \cdot \mathbf{v} = 0$ and $\nabla \cdot \mathbf{B} = 0$ and looked for situations where fast, steady and almost uniform reconnection would occur. There aim was to analyse the inflow region to determine how M_{Ai} varied with reconnection rate M_{Ae}.

They considered a small perturbation about a uniform external field $\mathbf{B}_e = B_e\mathbf{x}$ and used M_{Ae} as the implicit expansion parameter such that

$$\mathbf{B} = \mathbf{B}_e + \mathbf{B}_1 + \dots , \qquad \mathbf{v} = \mathbf{v}_1 + \dots , \qquad \text{and} \qquad p = p_0 + p_1 + \dots .$$

Equations (23) and (24) imply that $\mathbf{E} = \text{constant} = (0, 0, -E)$, so to first order the equations can be written

$$0 = -\nabla p_1 + (\nabla \times \mathbf{B}_1) \times \mathbf{B}_e/\mu_0 ,$$
$$\mathbf{E} + \mathbf{v}_1 \times \mathbf{B}_e = 0 .$$

$\nabla \cdot \mathbf{B}_1 = 0$ can be satisfied by writing the field as the curl of the flux function, $\mathbf{B}_1 = \nabla \times \mathbf{A}_1$. Thus, if we assume $\nabla p_1 \approx 0$,

$$-\frac{B_e}{\mu_0} \left(\frac{\partial^2 A_1}{\partial x^2} + \frac{\partial^2 A_1}{\partial y^2} \right) = 0 ,$$

which implies

$$\nabla^2 A_1 = 0 . \tag{25}$$

Separable solutions were sort for A_1,

$$A_1 = X(x)Y(y) ,$$

under the following boundary conditions,

$$\frac{\partial B_{1y}}{\partial x} = 0 , \qquad \text{on } |x| = L_e ,$$

$$B_{1x} = 0 , \qquad \text{on } |y| = L_e ,$$

$$B_{1y} = f(x) = \begin{cases} -2B_n & -L_e < x < -L \\ 2B_n x/L & -L < x < L \\ 2B_n & L < x < L_e \end{cases} , \qquad \text{on } y = 0 .$$

The first is a free-floating boundary condition which ensures that B_{1y} takes a maximum or minimum on the boundary and the final condition is an attempt to simulate the effect of the diffusion region and the shock (Figure 16).

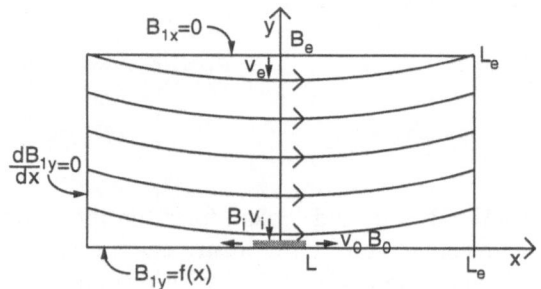

Fig. 16. Sketch of the magnetic field configuration in an almost-uniform steady-state reconnection situation [10]

The solution to this problems is

$$A_1 = -\sum_{m=0}^{\infty} \frac{2L_e a_m}{(2m+1)\pi} \cos\left[\frac{(2m+1)\pi}{2L_e}x\right] \cosh\left[\frac{(2m+1)\pi}{2L_e}(L_e - y)\right] ,$$

and so the components of the magnetic field are

$$B_{1x} = -\sum_{m=0}^{\infty} a_m \cos\left[\frac{(2m+1)\pi}{2L_e}x\right] \sinh\left[\frac{(2m+1)\pi}{2L_e}(L_e - y)\right] ,$$

$$B_{1y} = \sum_{m=0}^{\infty} a_m \sin\left[\frac{(2m+1)\pi}{2L_e}x\right] \cosh\left[\frac{(2m+1)\pi}{2L_e}(L_e - y)\right] ,$$

where

$$a_m = \frac{16 B_n \sin\left[(2m + 1)\,\pi L/2L_e\right]}{L/L_e\,(2m+1)^2\,\pi^2 \cosh\left[(2m+1)\,\pi/2\right]} \; .$$

This represents a Petschek type solution with a weak fast-mode expansion: the first order flow is uniform ($\mathbf{v}_1 = E_1/B_e \mathbf{y}$), but the second order flow is converging. The reconnection rate can be calculated from (17) and (21). It shows that as M_{Ai} increases so too does M_{Ae} until it saturates as Petschek anticipated.

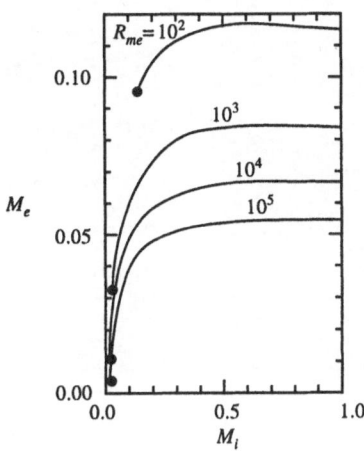

Fig. 17. A plot of M_{Ae} versus M_{Ai} for different R_{me} [10]

However, if we now include a pressure gradient then (25) becomes

$$\nabla^2 A_1 = -\frac{\mu_0}{B_e}\frac{dp_1}{dy} \; .$$

Using the same boundary conditions this leads to the new solution,

$$A_1 = -\sum_{m=0}^{\infty} c_m \left(b + \cos\left[\frac{(2m+1)\,\pi}{2L_e}x\right]\right) \cosh\left[\frac{(2m+1)\,\pi}{2L_e}(L_e - y)\right] \; ,$$

where $c_m = 2L_e a_m/((2m+1)\pi)$ and the new parameter b (relating to pressure) has a dramatic affect as it produces a whole range of different regimes:

- $b = 0$ Petschek's regime
- $b = 1$ Sonnerup-like solution: weak slow-mode expansion across whole inflow region
- $b < 0$ streamlines near y-axis are converging so tend to compress the plasma: slow-mode compressions
- $b > 1$ streamlines near y-axis are diverging so tend to expand the plasma: slow-mode expansions - "flux pile up"

• $0 < b < 1$ Hybrid family of slow and fast-mode expansions

The magnetic field structures of almost-uniform reconnection regimes are shown in Figure 18 with the different types of inflow indicated by dashed lines and the magnetic field indicated by solid lines. A plot of M_{Ae} versus M_{Ai} for almost-uniform reconnection regimes reveals that depending on the parameter b the rate of reconnection can vary greatly from a fast regime to a slow regime (Figure 18g).

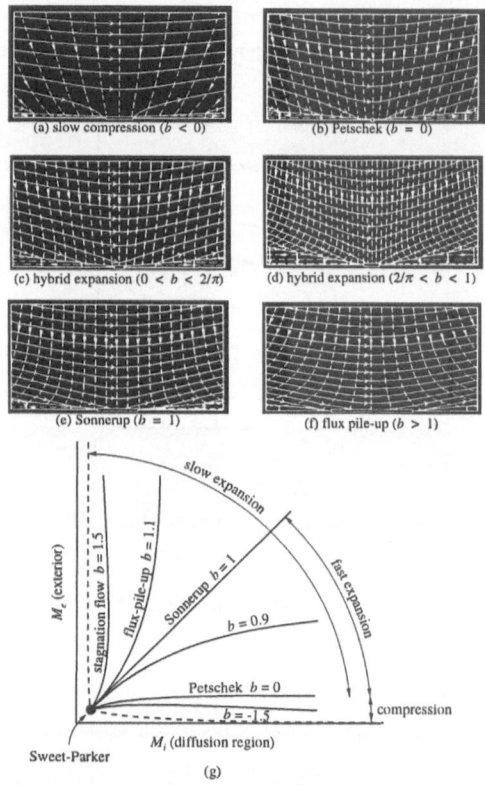

Fig. 18. Magnetic field configurations of almost-uniform steady-state reconnection situations with the pressure parameter b equal to (a) $b < 0$ (b) $b = 0$ (c) $0 < b < 2/\pi$ (d) $2/\pi < b < 1$ (e) $b = 1$ and (f) $b > 1$ [10]. (g) A plot of M_{Ae} versus M_{Ai} for different values of b

5.8   Numerical Experiments

All the reconnection regimes we have discussed so far have been analytical, but a considerable amount of work has been carried out numerically. Biskamp [9] attempted to reproduce Petschek reconnection numerically, but failed. He could only achieve a slow reconnection rate proportional to the Sweet-Parker rate and also found that the length and width of the current sheet increased as M_{Ae} or R_{me} increased (in total contrast to Petschek). He therefore concluded that Petschek reconnection (and therefore fast reconnection) did not exist for high R_{me}. This of course was very worrying to the solar physics community, who at the time only knew of Sweet-Parker and Petschek reconnection, because it implied that reconnection could not explain solar flares. Furthermore, his results showed some interesting features not explained by either of the two analytical reconnection models. These were strong jets observed along the separatrices and reverse current spikes observed at the end of the diffusion regions.

These results prompted an argument that still continues today as to whether fast Petschek reconnection exits. However, Biskamps boundary conditions were completely different from Petscheks. For example, the nature of inflow varied from converging to highly diverging, whereas Petschek had weakly converging inflow and the inflow fieldlines were highly curved with a large shock angle as opposed to almost uniform and a small shock angle in Petscheks models. These changes were enough to completely alter the nature of the reconnection as demonstrated by the almost-uniform family of reconnection regimes [10]. Furthermore, numerical experiments have been conducted by Priest and Lee [11] using boundary conditions consistent with Petschek's model where fast reconnection has been obtained.

6   3D Magnetic Reconnection

In 2D, reconnection has to take place at a null point and is defined as the transfer of flux across separatrices or as the process by which magnetic field topologies are changes. However, these definitions are not robust in 3D. Indeed, a neutral point in 2D is equivalent to a null-line in 3D which is structurally unstable. The definition of reconnection in 3D is still under debate partly because to define such a process we need to understand how the process works, and the field of 3D reconnection is still fairly new [13,14,15]. Of, course, reconnection is still a non-ideal process, however, it may occur both at neutral points and without neutral points in 3D.

6.1   3D Neutral Points

Structure

Before we can investigate 3D reconnection at neutral points we need to understand the topology of 3D neutral points. As in 2D, a 3D neutral point is a point at which all components of the magnetic field are equal to zero, $\mathbf{B} = \mathbf{0}$ and, as before, the local field near such a point can be written

$$\mathbf{B} = \mathbf{M} \cdot \mathbf{r} ,$$

where M is a 3x3 matrix and $\mathbf{r}$ is the position vector. The simplest form for M that allows for all the different types of 3D neutral points, but does not allow for duplication through rotation [19], is

$$
\begin{bmatrix}
1 & (q - j_\parallel)/2 & 0 \\
(q + j_\parallel)/2 & p & 0 \\
0 & j_\perp & -(p+1)
\end{bmatrix} ,
$$

where the current equals $\mathbf{j} = (j_\perp, 0, j_\parallel)$. Let the eigenvalues of M be denoted by λ_i and $\mathbf{x}_i$ be their corresponding eigenvectors. As in 2D, we find that satisfying the condition $\nabla \cdot \mathbf{B} = 0$ implies that the sum of the eigenvalues of M is equal to zero,

$$
\sum_{i=0}^{3} \lambda_i = 0 ,
$$

hence, λ_1 and λ_2 will be of one sign and λ_3 will be of the opposite sign.

If the two eigenvalues of the same sign are positive then the null is known as a positive null (or A-type [22]). If the eigenvalues of the same sign are negative then the null is known as a negative null (or B-type [22]).

Assuming that all the eigenvalues are real and distinct the equation for a fieldline can be written

$$
\mathbf{r}(k) = Ae^{\lambda_1 k}\mathbf{x}_1 + Be^{\lambda_2 k}\mathbf{x}_2 + Ce^{\lambda_3 k}\mathbf{x}_3 .
$$

If λ_1 and $\lambda_2 > 0$ and $\lambda_3 < 0$ then as

$$
k \to \infty \qquad \mathbf{r}(k) \to Ae^{\lambda_1 k}\mathbf{x}_1 + Be^{\lambda_2 k}\mathbf{x}_2 .
$$

This means that fieldlines that pass very close to the null will lie in surfaces parallel to the surface defined by eigenvectors $\mathbf{x}_1$ and $\mathbf{x}_2$ when they are directed away from the null. This surface is known as the fan plane (Figure 19a). However, as

$$
k \to -\infty \qquad \mathbf{r}(k) \to ce^{\lambda_3 k}\mathbf{x}_3 ,
$$

so the fieldlines that pass close to the null run parallel to the eigenvector $\mathbf{x}_3$ when they are directed towards the null. This line is known as the spine of the null (Figure 19a).

In all the 3D null configurations we consider here the spine is always taken to be directed along the z-axis. The components of the current $j_\perp$ refer to the current in the direction perpendicular to the spine, whilst $j_\parallel$ refers to that parallel to the spine.

Configurations

Clearly in 3D there are many more different types of neutral point configurations than just the four in 2D. Indeed, in 3D there are a whole class of

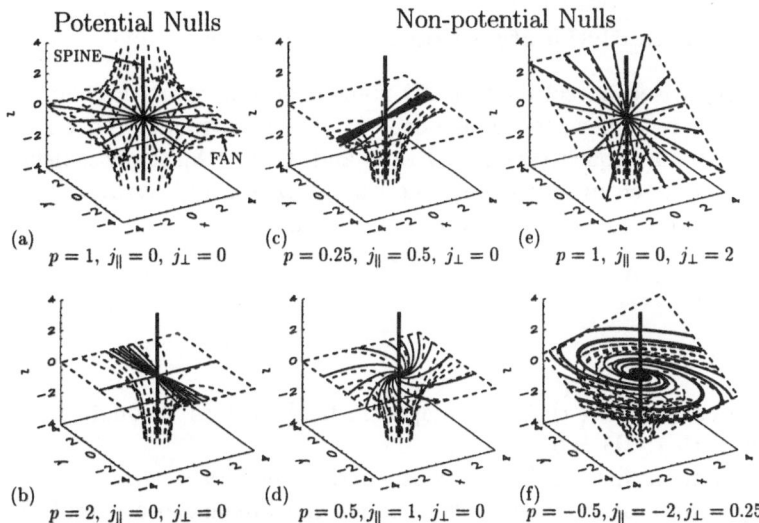

Fig. 19. Linear three-dimensional nulls. Potential nulls: (a) radial and (b) improper. Non-potential nulls: (c) improper and (d) spiral, both with only parallel current. (e) radial, with only perpendicular current and (f) spiral, with both component of current.

potential nulls whereas in 2D there is just one. The 3D potential nulls have the mathematical form

$$\mathbf{B} = (x, py, -(p+1)z) \, .$$

Their spine and fan planes are always at right angles to one another and the fieldlines in the fan plane are either radial (Figure 19a) or symmetrically gather in two bundles around the x or y axis (Figure 19b).

There are also many types of non-potential nulls. For example, a null with a field of the form

$$\mathbf{B} = (x + (q - j_\parallel)y/2, (q + j_\parallel)y/2 + py, -(p+1)z) \, ,$$

has only current along the spine, assuming that the spine is along the z axis. In such nulls the spine and fan are perpendicular, as in the potential case, but the fieldlines in the fan plane are either skewed, if $|j_\parallel| < q$ (Figure 19c), or spiraled, if $|j_\parallel| > q$ (Figure 19d). In the case where there is only a component of current perpendicular to the spine,

$$\mathbf{B} = (x, py, j_\perp y - (p+1)z) \, ,$$

assuming the spine is in the z direction. Such nulls have their fan plane inclined at some angle to the spine as seen in Figure 19e. Finally, nulls with both parallel and perpendicular components of current will have inclined fan planes and skewed or spiraled fieldlines (Figure 19f).

6.2   3D Kinematic Neutral Point Reconnection

Having familarised ourselves with 3D neutral points let us now move on to consider how reconnection could occur about these points. To do this we will follow the approach taken by Priest and Titov [18]. They considered a kinematic approach in much the same way as Green and Syrovatsky did for 2D reconnection, thus the equations considered are,

$$\mathbf{E} + \mathbf{v} \times \mathbf{B} = 0 , \tag{26}$$

$$\nabla \times \mathbf{E} = 0 . \tag{27}$$

under the assumption that the magnetic field evolves through a series of equilibria with $\mathbf{j} \times \mathbf{B} = 0$.

The magnetic field $\mathbf{B}$ is prescribed and a velocity $\mathbf{v}$ is imposed on the boundary. The system then evolves under the kinematic assumptions and the flow is investigated to see if it becomes singular at any point. If it does then it is taken to imply that one of our assumptions is wrong and that non-ideal effects are important in the region of the singularity. This region is therefore likely to be where the reconnection occurs.

Spine Reconnection

Let us consider the magnetic field $\mathbf{B} = (r, 0, -2z)$ and let us try and find a velocity field of the form $\mathbf{v}(r, \phi, z) = (v_r, 0, v_z)$ which will cause flow across the fan plane which lies in the xy plane.

Putting $\mathbf{B}$ and $\mathbf{v}$ into Ohms law implies

$$\mathbf{E} = (0, E, 0) ,$$

and so from steady state Faraday's law (27) we find that

$$\frac{\partial E}{\partial z} = 0 \qquad \text{and} \qquad \frac{1}{r} \frac{\partial Er}{\partial r} = 0 .$$

So

$$E = \frac{E_0(\phi)}{r} .$$

Now considering (26) $\times \mathbf{B}$ we can calculate the perpendicular component of the magnetic field,

$$\mathbf{v}_\perp = \frac{\mathbf{E} \times \mathbf{B}}{B^2} = \left(\frac{2E_0 z}{r(r^2 + 4z^2)}, 0, \frac{E_0}{r^2 + 4z^2} \right) .$$

Clearly the magnitude of $\mathbf{v}_\perp = E_0/B = E_0/r(r^2 + 4z^2)^{\frac{1}{2}}$ and so $\mathbf{v}_\perp$ is singular at origin and all the way along the z-axis, the spine of the null.

To see what effect such a flow would have on the magnetic field let us look at a flux surface and see how that evolves. Firstly we know that the equations for the fieldlines are $r^2 z = $ constant and $\phi = $ constant and flux surfaces are envelopes of these fieldlines. Suppose we consider the flux surface that intersects the cylindrical surface $r = 1$ at height $z = 1 - t \sin \phi$. Then for $0 < \phi < \pi$ the fieldlines move down at a speed $\dot{z} = -\sin \phi$ and trace out the curves shown in Figure 20a.

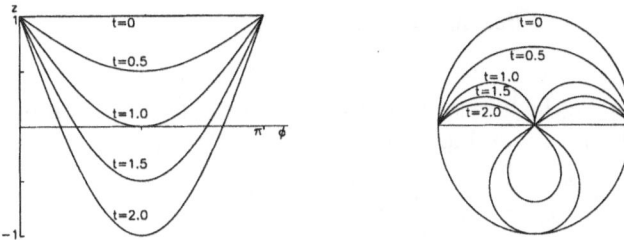

Fig. 20. The intersection of a flux surface near a radial null with (a) the $r = 1$ surface and (b) the $z = 1$ plane due to a flow across the fan plane at times $t = 0, 0.5, 1, 1.5$ and 2.

The flux surface through them is the surface $r^2 z = 1 - t \sin \phi$ which meets the $z = 1$ plane in the curve,

$$r^2 = 1 - t \sin \phi \qquad (t < 1/\sin \phi, \ 0 < \phi < \pi).$$

Footpoints with $\pi < \phi < 2\pi$ that cut the $z = -1$ plane will thread the $r = 1$ cylinder at $z = -1 - t \sin \phi$. The flux surface through them is $r^2 z = -1 - t \sin \phi$ and meets the $z = 1$ plane at

$$r^2 = -1 - t \sin \phi \qquad (-t \sin \phi > -1, \ |\phi - 3\pi/2| < -1/\sin \phi).$$

A sketch of the curves on $z = 1$ plane are shown in Figure 20b. The flux surfaces described evolve in the manner shown in Figure 21. So far we have the form for the magnetic field, the electric field and the perpendicular velocity, however, is it possible to calculate the total velocity $\mathbf{v}$? Substituting $\mathbf{B}$ and $\mathbf{E}$ into Ohms law we find

$$\frac{E_0}{r} - 2z v_r - r v_z = 0 \,,$$

therefore the velocity can be written,

$$\mathbf{v} = \left(\frac{E_0 - r^2 v_z}{2rz}, 0, v_z \right) \,.$$

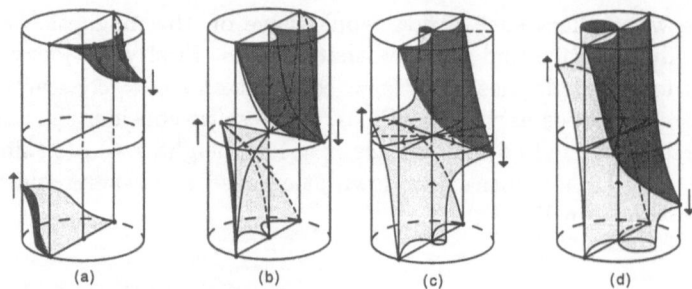

Fig. 21. Motions of flux surfaces in spine reconnection [18]

If we now make the further assumption that the flow is incompressible, $\nabla \cdot \mathbf{v} = 0$, we can find a separable solution for the velocity,

$$\mathbf{v} = \left(\frac{E_0 r^{2\alpha} - Az^\alpha}{2zr^{2\alpha+1}}, 0, \frac{Az^\alpha}{r^{2\alpha+2}} \right) ,$$

where α is an arbitrary constant and $\mathbf{v}$ is always singular at $r = 0$.

Fan Reconnection

Now consider the magnetic field $\mathbf{B} = (x, y, -2z)$ and look for solutions that will give us continuous motions on the surfaces $z = \pm 1$, i.e. look for solutions that drive in parallel lines across $z = \pm 1$ planes. Such a solution is given by

$$\mathbf{E} = \frac{1}{(4 + y^2 z)^{3/2} z^{1/2}} \left((4 + y^2 z)z, -xyz^2, 2x \right) ,$$

and

$$\mathbf{v}_\perp = \frac{1}{w} \left(-2xy(z^3 - 1), -2(x^2 + 4z^2 + y^2 z^3), -yz \right) ,$$

where $w = (x^2 + y^2 + 4z^2)(4 + y^2 z)^{3/2} z^{1/2}$. A $z \to 0$, $v_{\perp x}$ and $v_{\perp y} \to \infty$ like $z^{-1/2}$. So steady ideal MHD breaks down in the fan plane. On $z = 1$, $v_{\perp x} = 0$ and $v_{\perp y}$ are independent of x, thus the footpoints of the fieldlines that thread $z = 1$ are driven at a constant speed parallel to the y-axis. On the cylinder $x^2 + y^2 = 1$ the footpoints of the fieldlines swing around and down the cylinder when $x, z > 0$, and around and up cylinder when $x, z < 0$. This means that the flux surface in $x = 0$ plane reconnects at the fan. A sketch of the evolution of the flux surfaces during fan reconnection is shown in Figure 22.

6.3   3D Numerical Experiments

These kinematic results have been backed up by numerical experiments [23,24]. Rickard and Titov [23] solved the linearized resistive MHD equations under

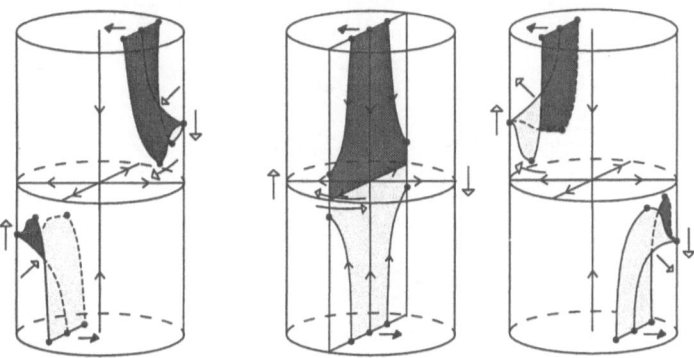

Fig. 22. Motions of flux surfaces in fan reconnection [18]

the line tied boundary condition. The initial magnetic field was taken to be that of a radial null, $\mathbf{B} = (B_r, B_\theta, B_z) = (r, 0, -2z)$, which was perturbed by one of two types of perturbation, either an $m = 0$ or $m = 1$ (Figure 23). An $m = 0$ perturbation is one in which the spine is either compressed or stretched or the null is rotated about the spine. In all $m = 0$ cases the spine and fan maintain their right-angled inclination. $m = 1$ perturbations involve a bending of the spine or fan such that the inclination between the spine and fan increases or decreases.

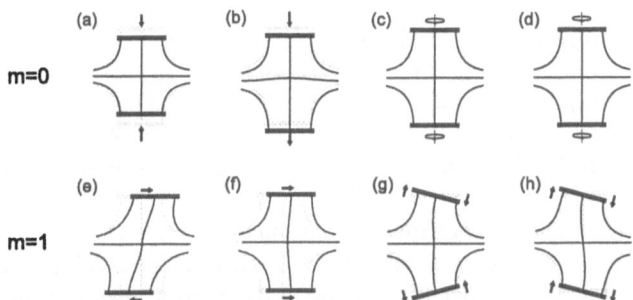

Fig. 23. Typical perturbations that may be suffered by a 3D null. In these diagrams the spine is vertical and the fan is horizontal [23]

In the first case the null is twisted about the spine first one way and then the other. This led to a the generation of a B_θ that propagated in towards the null along the fan plane (Figure 24a and 24b). A corresponding j_z component of current similarly ran in along the fan plane then spread out along the spine implying the likelihood that the reconnection will occur along the spine (Figure 24c and 24d).

In the second case the null under went a small $m = 1$ perturbation which led initially to a global current build up. After a few alfven crossing times,

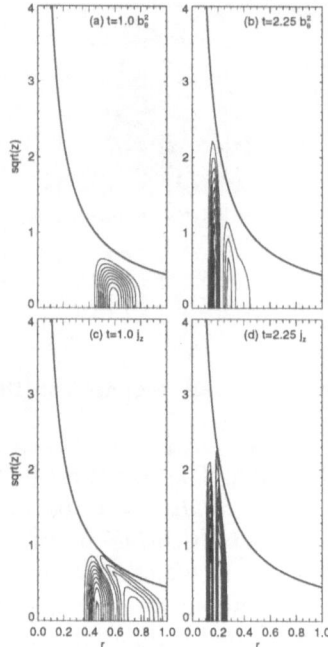

Fig. 24. The resulting B_θ and j_z at times $t = 1$ and $t = 2.25$ induced in a radial null due to an $m = 0$ perturbation.

though, the current dissipated such the j_r and j_θ components spread out along the fan plane whilst the j_z component lay parallel to the spine of the null (Figure 25a). However the j_θ and j_z components of current rapidly decay after a few alfven times whilst the j_r component tends to a finite value. This indicates the likelihood that reconnection would take place in the fan plane (Figure 25b).

Of course, in their experiments they just used a linearised resistive code, however, the same sort of results have been obtained using a full resistive MHD code [24]. Galsgaard [24] in his experiments also assumed line tying at the boundaries and choose his initial field to be a radial null. As in the previous work his results show that an $m = 1$ perturbation leads to current accumulations in the fan plane.

6.4   Reconnection Without Nulls: Quasi-separatrix Layers

So far we have just discussed reconnection in three-dimensions at neutral points, however, reconnection can also occur without neutral points. Priest and Demoulin [17] attempted a kinematic analysis of this type of reconnection.

They considered the magnetic field

$$\mathbf{B} = (x, -y, l), \qquad l \ll 1,$$

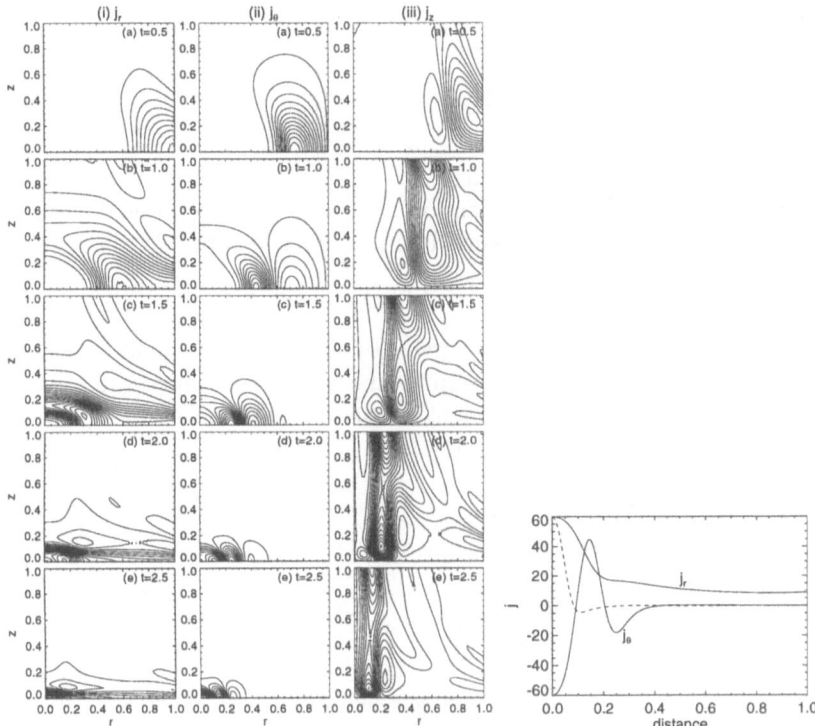

Fig. 25. (a)The resulting j_r, j_θ and j_z at times $t = 0.5$, 1, 1.5, 2 and $t = 2.5$ induced in a radial null due to an $m = 1$ perturbation. (b)The magnitude of j_r, j_θ and j_z are plotted against time [23]

which clearly has no points where $\mathbf{B} = 0$. The fieldlines of such a field are defined by

$$x = x_0 e^{z/l} \quad \text{and} \quad y = y_0 e^{-z/l} .$$

Let the footpoint of one fieldline in the $z = 0$ plane be $(x_0, y_0, 0)$. This fieldline will thread the $z = 1$ plane at the point $(x_0 e^{1/l}, y_0 e^{-1/l}, 1)$. Now let us assume that the footpoint of this fieldline moves from $(x_0, y_0, 0)$ to $(-x_0, y_0, 0)$, thus the corresponding endpoint in the $z = 1$ plane would move from $(x_0 e^{1/l}, y_0 e^{-1/l}, 1)$ to $(-x_0 e^{1/l}, y_0 e^{-1/l}, 1)$. Clearly the footpoint on $z = 1$ plane moved a much greater distance and therefore faster than the footpoint on $z = 0$ plane. It in fact moves a factor $e^{1/l}$ times faster and since $l << 1$, $e^{1/l}$ is very large, thus the fieldlines either moved unphysically fast or reconnected.

To determine the region in which the fieldlines have this type of behaviour and hence which fieldlines reconnect they investigated the gradient of the displacement of the fieldlines given by the displacement gradient tensor (F)

which equals

$$F = \begin{pmatrix} \partial x_1/\partial x_0 & \partial x_1/\partial y_0 \\ \partial y_1/\partial x_0 & \partial y_1/\partial y_0 \end{pmatrix}.$$

If F is very large in some region then that region is said to be a quasi-separatrix layer in which reconnection can occur. For example, consider the above magnetic field, but this time bound it by the cube

$$|x| \leq 1/2, \quad |y| \leq 1/2 \quad \text{and} \quad 0 \leq z \leq 1.$$

If the fieldline footpoints are given by $(x_0, y_0, 0)$, as before, and $x_0 < e^{-1/l}/2$ then the fieldlines will map to the upper surface of the cube since $x_1 < 1/2$. The displacement gradient tensor therefore becomes

$$F = \begin{pmatrix} e^{1/l} & 0 \\ 0 & e^{-1/l} \end{pmatrix}.$$

The magnitude (or norm) of this tensor (N) is given by

$$N = \sqrt{\frac{\partial x_1}{\partial x_0} + \frac{\partial x_1}{\partial y_0} + \frac{\partial y_1}{\partial x_0} + \frac{\partial y_1}{\partial y_0}} = \sqrt{e^{2/l} + e^{-2/l}} \approx e^{1/l}.$$

However, if $e^{-1/l}/2 < x_0 < 1/2$ then the fieldlines map to the side of the box and the displacement gradient tensor is

$$F = \begin{pmatrix} 0 & 0 \\ 2y_0 & 2x_0 \end{pmatrix}.$$

So in this case the norm of the tensor equals

$$N = 2\sqrt{x_0^2 + y_0^2}.$$

Therefore, as x_0 increases from 0 to $e^{-1/l}/2$ the other end of the fieldline rapidly increases from 0 to $1/2$ whilst y_1 remains equal to $e^{-1/l}y_0$. And since $l \ll 1$ and $N \approx e^{1/l}$ the norm of the tensor is large. However, as x_0 increases from $e^{-1/l}/2$ to $1/2$, x_1 remains constant at $1/2$ and the norm of the tensor $N < \sqrt{2}$ so is small. Therefore regions parallel to y-axis out to $|x| = e^{-1/l}$ are where reconnection takes place. These regions are called quasi-separatrix layers (QSLs). A sketch of reconnection at quasi-separatrix layers is shown in Figure 26. Currently experiments are underway to determine if this type of behaviour can be reproduced numerically.

7   Summary

This review gives an overview to the basics of magnetic reconnection and the classical aspects of reconnection in both 2D and 3D. In particular, we have seen that,

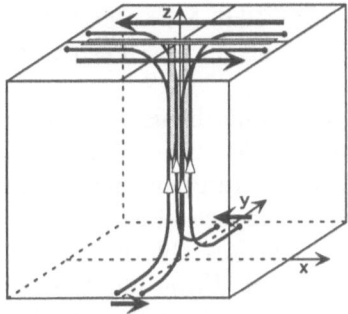

Fig. 26. A sketch of the motion of fieldlines near a quasi-separatrix layer [17]. The shaded regions represent the quasi-separatrix layers.

- Reconnection is an efficient means of converting magnetic energy to thermal and bulk kinetic energy
- Reconnection only occurs locally in the solar corona where short length scales exist
- Magnetic fields evolves due to advection and diffusion
- The magnetic Reynolds number R_m is $\approx 10^8 - 10^{12}$ in the corona and so, in general, the magnetic field is frozen-in to plasma
- 2D Reconnection only occurs at neutral points
- Steady-state reconnection can be fast or slow depending on boundary conditions
  - ⋆ Sweet-Parker reconnection is slow
  - ⋆ Petschek reconnection is fast
- Petschek reconnection is not special but just one of many fast reconnection regimes
- Petschek reconnection can be obtained from numerical experiments
- Reconnection in 3D is quite different from reconnection in 2D since it can occur:
  - ⋆ at neutral points
  - ⋆ without neutral points
- Reconnection at neutral points is either
  - ⋆ spine reconnection - reconnection of fieldlines across the spine
  - ⋆ fan reconnection - reconnection of fieldlines across the fan plane
- Reconnection without neutral points occurs in quasi-separatrix layers

References

1. Giovanelli, R.G. (1946) A Theory of Chromospheric Flares. Nature 158:81-82
2. Cowling, T.G. (1953) In: Kuiper G.P (Ed.) The Sun. University of Chicago Press
3. Dungey, J.W. (1953) Conditions for the Occurrence of Electrical Discharges

in Astrophysical Systems. Phil. Mag. Ser. 7 44:725-738

4. Sweet, P.A. (1958) The Neutral Point Theory of Solar Flares. IAU Symp. 6:123-134

5. Parker, E.N. (1957) Sweet's Mechanism for Merging Magnetic Fields in Conducting Fluids. J. Geophys. Res. 62:509-520

6. Parker, E.N. (1963) Solar Flare Phenomena and the Theory of Reconnection and Annihilation of Magnetic Fields. Phys. Rev. 999:177-211

7. Furth, H.P., Killeen, J., and Rosenbluth, M.N. (1963) Finite Resistive Instabilities of a Sheet Pinch. Phys. Fluids 6:459

8. Petschek, H.E. (1964) Magnetic Field Annihilation. AAS-NASA Symp. on Phys. of Solar Flares. NASA SP 50:425-439

9. Biskamp, D. (1986) Magnetic Reconnection via Current Sheets. Phys. Fluids 29:1520-1531

10. Priest, E.R. and Forbes, T.G. (1986) New Models for Fast Steady State Reconnection. J. Geophys. Res. 91:5579-5588

11. Priest, E.R. and Lee, L.C. (1990) Nonlinear Magnetic Reconnection Models with Separatrix Jets. J. Plasma Phys. 44:337-360

12. Scholar, M. (1989) Undriven Magnetic Reconnection in an Isolated Current Sheet. J. Geophys. Res. 94:8805-8812

13. Schindler, K., Hesse, M. and Birn, J. (1988) General Magnetic Reconnection, Parallel Electric Fields and Helicity. J. Geophys. Res. 93:5547-5557

14. Hesse, M. and Schindler, K. (1988) A Theoretical Foundation of General Magnetic Reconnection. J. Geophys. Res. 93:5559-5567

15. Hornig, G. and Schindler, K. (1996) Magnetic Topology and the Problem of its Invariant Definition. Phys. Plasmas 3:781-791

16. Lau,Y.T. and Finn, J.M. (1990) Three-dimensional Kinematic Reconnection in the Presence of Field Nulls and Closed Field Lines. Astrophys.J. 350:672-691

17. Priest, E.R. and Démoulin, P. (1995) Three-dimensional Magnetic Reconnection without Null Points 1. Basic Theory of Magnetic Flipping. J. Geophys. Res. 100:443-463

18. Priest, E.R. and Titov, V.S., (1996) Magnetic Reconnection at Three-dimensional Null Points. Phil. Trans. R. Soc. Lond. 354:2951-2992

19. Parnell, C.E., Smith, J.M., Neukirch, T. and Priest, E.R. (1996) The Structure of Three-dimensional Magnetic Neutral Points. Phys. Plasmas 3:759-770

20. Green, R.M. (1965) Models of Annihilation and Reconnection of Magnetic Fields. IAU Symp. 2:398-404

21. Syrovatsky, S.I. (1971) Formation of Current Sheets in a Plasma with a Frozen-in Strong Magnetic Field. Soviet Phys. JETP 33:933-940

22. Cowley, S. (1973) A Qualitative Study of the Region Between the Earths Magnetic Field and an Interplanetary Field of Arbitrary Orientation. Radio Science 8:903-913

23. Rickard, G.J. and Titov, V.S. (1996) Current Accumulation at a Three-

dimensional Magnetic Null. Ap. J. 472:840-852

24. Galsgaard, K., Rickard, G.J., Reddy, R.V. and Nordlund, A. (1996) Dynamical Properties of Single and Double 3D Null Points. In: Bentley, R.D., Mariska, J.T. (Eds) Magnetic Reconnection in the Solar Atmosphere A.S.P. CONF. SER. 111:82-88

Structuring of the Solar Plasma
by the Magnetic Field

Pascal Démoulin and Karl-Ludwig Klein

Observatoire de Paris, Section de Meudon, DASOP, UMR 8645 (CNRS),
F-92195 Meudon Cedex, France;
e-mail: Pascal.demoulin@obspm.fr, Ludwig.klein@obspm.fr

Abstract. This paper presents a simplified overview of the role of the magnetic field in the solar atmosphere. The magnetic field emanating from the solar interior governs energy transport and plasma motions in the outer solar atmosphere. Thereby it creates structure, such as coronal holes, loops and prominences, and the dynamical phenomena known as coronal mass ejections and flares. The magnetic field is also thought to be at the origin of the coronal heating, so of the corona itself. An overview of atmospheric structure is presented, followed by illustrations on present ideas on the interaction between plasma and magnetic field. The physical conditions in the corona are briefly compared to those in the magnetosphere. The emphasis is then put on the energetic processes from the largest ones (coronal mass ejections) over flares and X-ray bright points to coronal heating. In all cases magnetic reconnection is likely to play a key role. Solar prominences are then described because their observations provide important information on the surrounding coronal magnetic field. Finally the implications of processes in the convection zone on the physics of the corona and of the interplanetary medium are illustrated for the case of formation, storage and ejection of twisted magnetic flux tubes.

1 Introduction

Solar magnetic fields are created at the bottom of the convection zone from the kinetic energy of the dense plasma. Buoyant flux tubes rise into the atmosphere, such that the magnetic field eventually fills the corona, creates its structures and governs its dynamics. This provides energy transport into the atmosphere, where the interaction between the plasma and the magnetic field leads to dynamical phenomena including heating, particle acceleration and the ejection of material.

The term "atmosphere" designates the outer envelope of the Sun, where the electromagnetic radiation is generated. Section 2 of this review presents the basic temperature regimes of the atmosphere (photosphere, chromosphere, corona), outlines the different types of electromagnetic radiation, and illustrates the most important plasma structures in the (low-beta) corona that we interpret as being generated by the magnetic field.

Present ideas on the interaction of plasma and magnetic fields in the corona are discussed for selected topics in the following Sections. Emphasis is laid on the structure and evolution of the magnetic field configurations.

The physical system is always described by MHD equations (see e.g. Heyvaerts 2000). Each selected topic is treated in two basic parts: the first is general, introducing the subject, and the second focuses on one aspect of present research. Several textbooks treat these subjects with more detail: Stix (1991) provides an introduction to solar physics with emphasis on the interior and low atmosphere. Tutorial lectures and reviews on all fields of solar physics are presented in the books edited by Cox et al. (1991) and Schmelz and Brown (1992). Priest (1982) is a classical monograph providing both a summary of observations and models with emphasis on the solar atmosphere viewed from the MHD point of view. The solar corona has been recently reviewed by Golub and Pasachoff (1997). A review based mostly on white-light and radio sounding observations of large-scale properties is Bird and Edenhofer (1990). Flares and filaments have been described in detail by Tandberg-Hanssen and Emslie (1988) and Tandberg-Hanssen (1995), respectively. Coronal mass ejections are reviewed in the book edited by Crooker, Joselyn and Feynman (1997). For kinetic aspects of solar and stellar coronae, which are not discussed here, the reader is referred to Benz (1993).

The release of magnetic energy in the corona occurs over a broad range going from 10^{25} J within minutes to hours (large flares, coronal mass ejections), possibly down to tiny and yet undetectable events such as Parker's suspected nanoflares (10^{16} J). One main, but not the only, process invoked is magnetic reconnection, which has been extensively studied in 2-D (see e.g. Parnell 1998). This approach has highlighted the importance of magnetic topology (Sect.4.1). Recent works have evolved from 2D to 3D reconnection, giving a completely new view; this is shortly summarized to describe the energy release from the most spectacular events of large-scale magnetic restructuring (coronal mass ejections, in Sect.4.2) to small-scale phenomena including flares (Sect.4.3), X-ray bright points (Sect.4.4), and coronal heating (Sect.4.5). The prominences / filaments are then described in Sect.5 because of their particular plasma structuration and because they provide important information on the surrounding coronal magnetic field. Finally in Sect.6, an attempt is presented to link a part of the physics of the convective zone, of the corona and of the interplanetary medium.

2   The Structure of the Solar Atmosphere

2.1   General Aspects and Temperature Regimes

The schematic evolution of temperature with height in the solar atmosphere is plotted in Fig. 1. The *photosphere* is the deepest layer directly observable through its electromagnetic radiation (mainly visible light). The level of zero altitude ("bottom" of the photosphere) is defined as the height where the optical depth of radiation at 500 nm wavelength is unity. The temperature of the black body which gives the closest approach of the observed intensity spectrum of the radiation is ~5700 K, but local temperature depressions

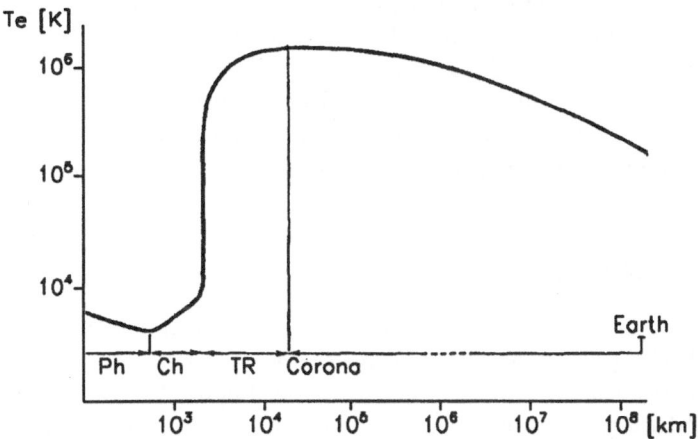

Fig. 1. Evolution of the electron temperature in the solar atmosphere as a function of height above the photosphere. The curve gives a schematic view of an "average" quiet atmosphere. Ph=photosphere, Ch=chromosphere, TR=transition region

(e.g. sunspots) and enhancements (faculae) by up to 1000 – 2000 K exist. Temperature first decreases with increasing altitude to a minimum of about 4400 K ("top" of the photosphere), over a distance of some hundreds of km. The cooler gas in the upper photosphere creates many absorption lines in visible light which are superposed upon the blackbody spectrum. Doppler and Zeeman effect measurements of these lines give the basic information on material flows and magnetic fields which govern the structure and dynamics of the overlaying atmosphere.

Above the photosphere temperature rises with increasing altitude. The region between the temperature minimum and $\sim$ 25 000 K is called the *chromosphere* because its optical and UV emission lines create a colored ring around the lunar disk during an eclipse. Besides in UV, the chromosphere can be observed in optical absorption lines (e.g. Hα) and at short radio wavelengths: the electron density in the atmosphere far from active regions is such that radio emission generated by free-free transitions (acceleration of free electrons by the electrostatic field of ions) can escape at frequencies above a few GHz (centimeter wavelengths). The thickness of the chromosphere is between one and a few thousand km.

The *corona* is the outer region of the atmosphere, with temperatures $\geq 10^6$ K. The narrow region between the upper chromosphere and the corona is called the (chromosphere-corona) *transition region*. The corona is visible in white light due to the scattering of photospheric light by free electrons ("K-corona") and dust ("F-corona"). Coronal emission comprises spectral lines of highly ionized heavy elements in visible, UV and EUV (especially Fe and Ca), as well as continuum emission (X, radio) due to free electrons. In the presence of strong magnetic fields, i.e. in the low corona above sunspots, centimetric cyclotron radio emission at low (2 to 4) harmonics of the electron cyclotron frequency (called also gyroresonance emission) is observed, while various types of transient anisotropic electron distribution functions in the active corona generate electromagnetic emission at centimetric and longer wavelengths via collective processes such as beam-plasma or loss-cone instabilities.

2.2   Magnetic Fields and the Structure of the Solar Atmosphere

Due to magnetic fields generated by a dynamo mechanism in the *convection zone* underneath the photosphere, the atmospheric structure, especially in the corona, is far from being spherically symmetric. Figure 2 shows views of different temperature regimes of the atmosphere through images at different wavelengths taken on a day of weak activity ("quiet" atmosphere, Fig. 2.a) and on a day when two active regions are present in the eastern hemisphere (left half of the images in Fig. 2.b, zoom on one of the active regions in Fig. 2.c).

Zeeman effect measurements in photospheric lines provide maps of the line-of-sight component of the magnetic field (top row) where active regions appear as magnetic field concentrations with a bipolar or a more complex multipolar configuration. Photospheric continuum emission (second row from top) shows sunspots, which are darker (cooler) than the average photosphere, and surrounding faculae, which are brighter (hotter) than the quiet photosphere at a given optical depth. Chromospheric lines (central row) show bright plages above regions of strong photospheric fields. The chromospheric lines also reveal structure outside active regions, such as the dark filaments (Fig. 2.a, near the south-eastern limb; Fig. 8a for a detailed view of another filament), which are concentrations of cool matter ($\sim$7000 K) within the hot ambient corona. When seen at the limb, these structures are bright with respect to the dark sky, and are called prominences for historical reasons (e.g. the gray feature embedded within the background corona at an angle $\sim$ 36 deg clockwise from top in Fig. 3).

X-ray (Fig. 2.a, second panel from bottom) and radio emission (bottom) of the quiet corona trace the thermal plasma, including the extended regions of low brightness (low electron density and temperature) at the poles and on the disk: coronal holes. These features have no evident counterpart in the underlying atmosphere. They overly photospheric regions with weak unipolar magnetic fields, and the coronal field lines are open towards interplanetary space. Coronal holes are the source of the fast solar wind. The corona of active regions consists of hot and dense plasma loops giving bright X-ray emission (Fig. 2.b, c). In these regions the plasma delineates predominantly (but not exclusively) closed magnetic field structures. The numerous bright points found in EUV- and X-ray images all over the Sun (Fig. 2.a) overly small bipolar magnetic field patterns in the photosphere outside active regions.

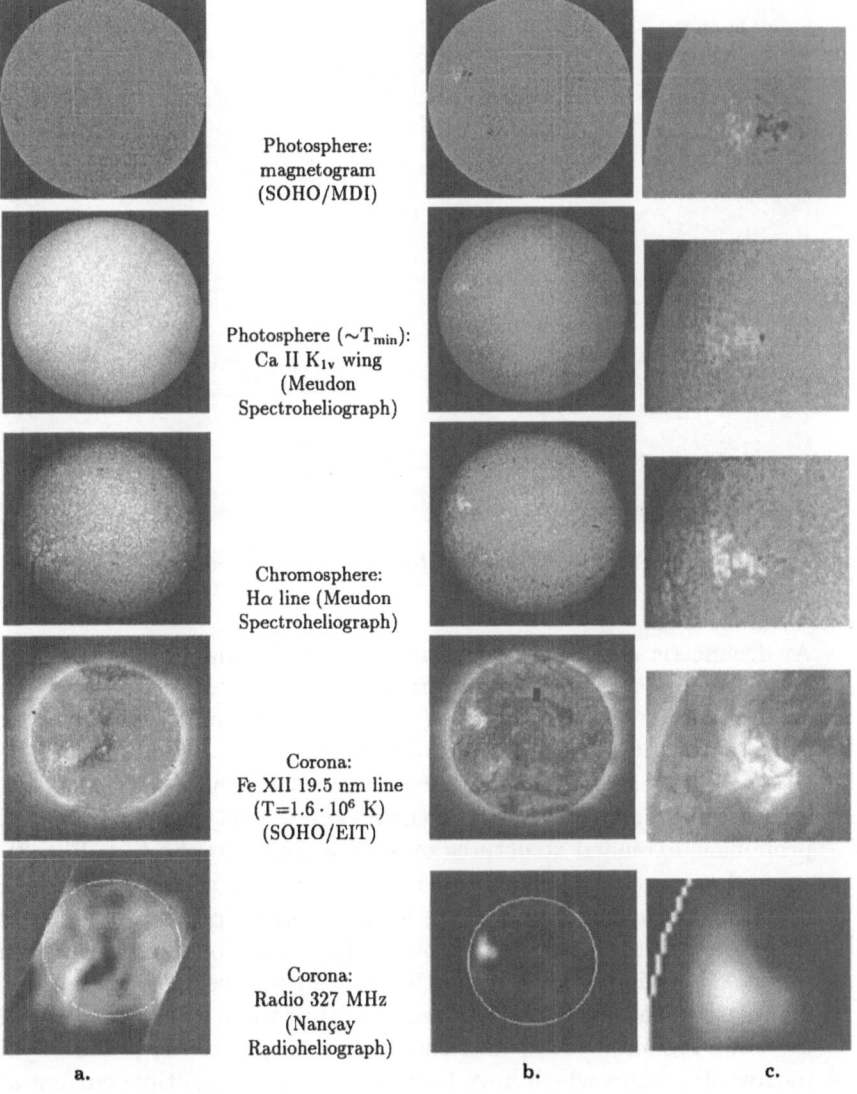

Fig. 2. Views of the solar atmosphere at different wavelengths, during a quiet period (a) and in the presence of active regions (b). (c) is a zoom of the north-eastern (upper left) active region

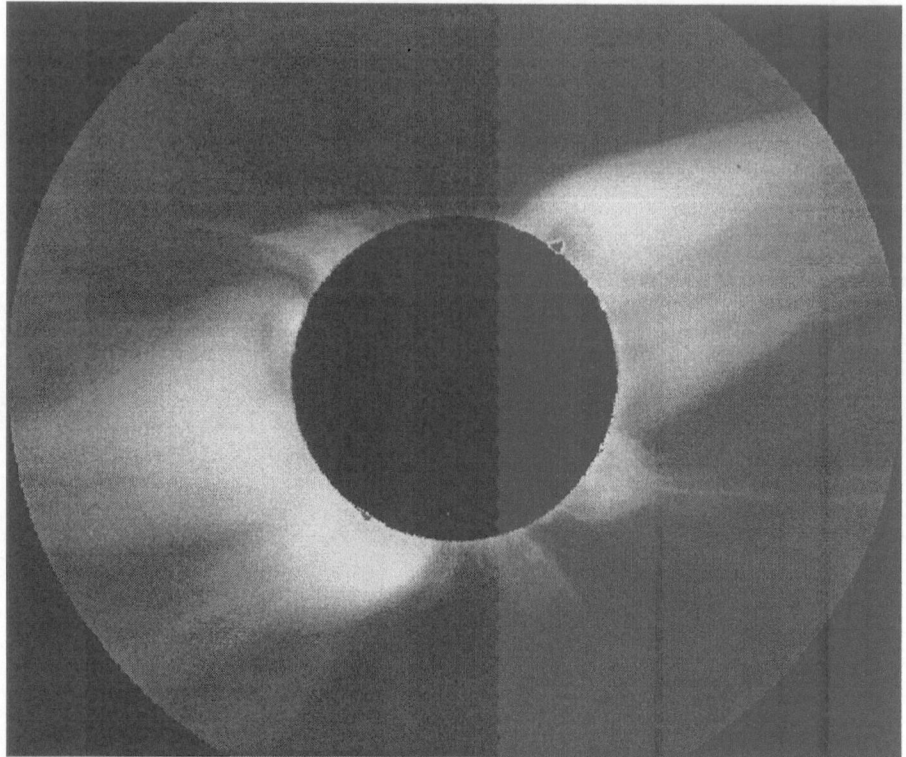

Fig. 3. Solar eclipse in white light (courtesy *High Altitude Observatory*, Boulder)

At decimetric and longer wavelengths suprathermal electrons accelerated in active regions provide a bright emission through collective processes ("noise storms") that outshine emission of the hot thermal plasma by one to two orders of magnitude (Fig. 2.b, c, bottom).

At heights above (1–2) $R_\odot$ above the photosphere the plasma structures seen in white light indicate that the magnetic field is mostly open. The most prominent extended structures on eclipse photographs (e.g. Fig. 3) are the coronal streamers. Their shape suggests that a closed magnetic field configuration at low heights, covering a large surface on the Sun, narrows with increasing altitude, and is eventually torn open by the outward streaming (slow) solar wind. Closer inspection of eclipse observations and of measurements using the signals of artificial or cosmic radio sources traveling through the outer regions of a streamer show numerous, more or less radially oriented, narrow structures which have been interpreted as multiple current sheets on top of the closed magnetic structure at the base of the streamer (Woo et al. 1995).

The atmospheric structures identified in these figures are quasi-steady in that they keep a similar shape for several days or weeks. But a permanent energy input is required to maintain the hot plasma and the nonthermal electron populations.

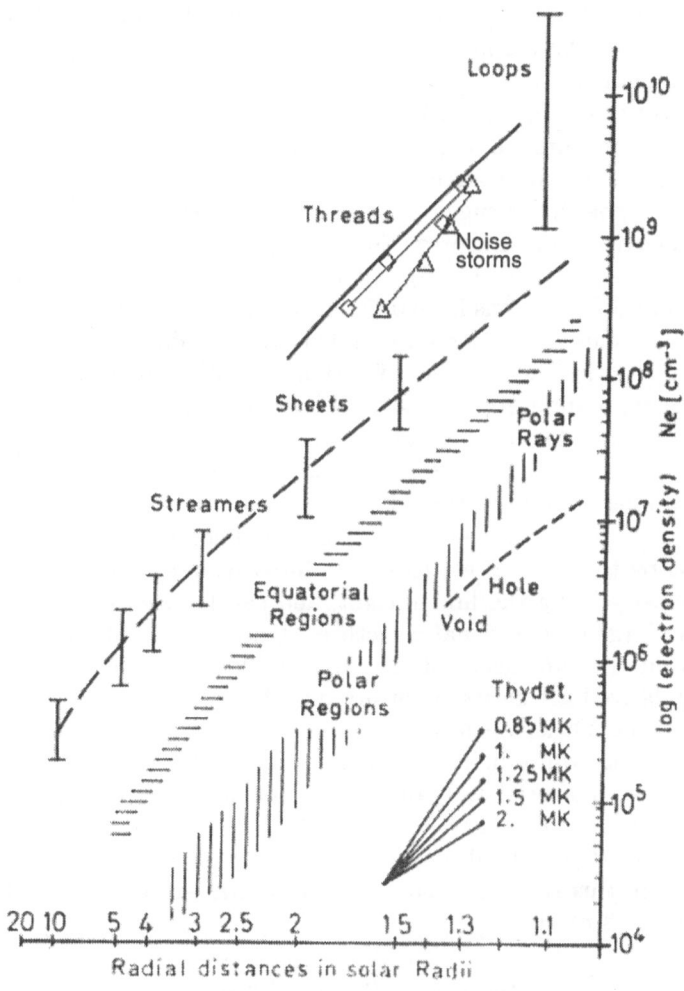

Fig. 4. Density vs. height profiles in different coronal structures (inferred from white light observations, Koutchmy 1994 and from noise storm observations with the *Nançay Radioheliograph*)

Coronal Electron Density The analysis of images in coronal emission lines (visible light, EUV, X) shows that the low corona has a heterogeneous (electron) temperature structure, ranging from the vicinity of 10^6K K to $\sim 7 \cdot 10^6$ K. Plasmas of different temperatures, seen in different wavelength ranges, may be confined in different loop structures. Furthermore the legs of identified loops are usually cooler than their summits. Commonly quoted values of the electron density at the top of soft X-ray loops in active regions are in the range $(3 - 13) \cdot 10^9$ cm^{-3} (e.g. Yoshida and Tsuneta 1996), and similar values are inferred from white-light observations (Fig. 4). These are lower limits, due to the assumption that the plasma fills a loop whose dimension along the line of sight is comparable to the observed width.

Ion temperatures can be derived from the widths of EUV emission lines, provided that competing broadening effects such as unresolved small-scale motions of the plasma in the source region can be corrected for. Recent SoHO analyses of coronal holes suggest that ion temperatures are considerably higher than electron temperatures (Tu et al. 1998; Wilhelm et al. 1998; David et al. 1998) - e.g. $T_i \simeq (1 - 5) \cdot 10^6$ K for Ne ions as opposed to $T_e < 10^6$ K, and that different ions have different temperatures. The decreasing role of collisions seems to become noticeable in the low-density plasma of coronal holes, even at altitudes as low as a fraction of a solar radius. The large ratio of ion-to-electron temperature in coronal holes is similar to the results of in situ measurements in fast solar wind streams by Helios at 0.3 AU, while in slow solar wind streams the electron temperature seems slightly higher than the proton temperature (Schwenn 1990).

Part of the coronal white-light emission is due to Thomson scattering of photospheric light by the coronal electrons. Its intensity measures the integrated electron density along the line of sight, and can be used to infer the electron density in large-scale coronal structures. Using eclipse observations and assumptions on the dimension of the sources along the line of sight, Koutchmy (1994) derived the electron density distributions of Fig. 4 for various coronal structures. Other density measurements for the quiet Sun and coronal holes from both white-light continuum and EUV lines give curves similar to those of the equatorial and polar regions of the quiet Sun in Fig. 4 (Withbroe 1988; first SoHO results in ESA SP-404). The plot represents log n_e as a function of $R_\odot/r$, the inverse of the heliocentric distance in units of the solar radius. In this representation the barometric isothermal density law in an unmagnetized plasma

$$n_e(r) = n_e(R_\odot) \exp[-\frac{R_\odot}{H_\odot}(1 - \frac{R_\odot}{r})]$$

is a straight line. Here $H_\odot = \frac{kT}{\mu m_p g_\odot}$ is the density scale height, k Boltzmann's constant, $g_\odot$ the gravitational acceleration at the solar surface, m_p the proton rest mass, $\mu = 0.6$ the mean molecular weight. $n_e(R_\odot)$ is the reference density at the base of the corona.

The local electron density can in principle be inferred from the measurement of positions of dm/m-wave noise storms (cf. Fig. 2.b,c bottom panels) without the need to inverse the line-of-sight integration. The radiation is generated close to the plasma frequency. The open diamonds and triangles labeled 'Noise storms' in Fig. 4 were derived from the position measurements at four frequencies of the centroids of two noise storms at the solar limb (data from the *Nançay Radioheliograph*). The straight lines give least-squares fits to the two data sets. The density profiles are quite similar to those inferred from the visible-light-observations in dense coronal structures.

The empirical density profiles at heliocentric distances $< 2\,R_\odot$ have slopes close to those predicted by the isothermal hydrostatic model with $T = (1-2) \cdot 10^6$ K, as indicated for different temperatures by the bundle of straight line segments in the lower right corner of Fig. 4. The absolute densities differ by up to three orders of magnitude between different coronal structures. While the agreement with a static model is not a surprise, given that the flow speeds at low altitudes are largely subsonic, it must be kept in mind that the measured densities are average values, due to the line-of-sight integration of the optical data and to the smearing out of the radio structures by coronal scattering and the antenna beam.

Magnetic Field Quantitative information on magnetic fields is mostly derived from Zeeman or Hanle effect measurements of spectral lines from rather cool plasmas, in the photosphere or in prominences. Coronal emission lines are too broad to allow Zeeman effect measurements. It is general use to infer the topology of the coronal magnetic field from the extrapolation of the measured photospheric field, using a current-free or a force-free model, and to constrain the models by the observed plasma structure. Direct determinations of the coronal magnetic field are only possible through the measurements of the circular polarization of thermal free-free emission at centimetric wavelengths and from observations of centimetric cyclotron emission in the coronal part of active regions above sunspots. The latter technique reveals that magnetic fields above 0.1 T exist in the $T \geq 10^6$ K plasma at several 10^3 km above sunspots in the photosphere (cf. reviews by Alissandrakis 1994; Klein 1992).

The magnetic field strength in the high corona, at altitudes between 2 and $9\,R_\odot$ above the photosphere, can be measured through the Faraday rotation of a linearly polarized signal from interplanetary spacecraft or from cosmic radio sources on its way through the corona (review by Bird and Edenhofer 1990). Measurements during weak solar activity show the field strength to decrease as the sum of an $(\frac{r}{R_\odot})^{-3}$ and an $(\frac{r}{R_\odot})^{-2}$ term at heights $> 2\,R_\odot$ above the photosphere. This also fits in situ measurements at 0.3 AU.

Plasma Parameters in Solar Atmospheric Structures A summary of typical plasma parameters in different atmospheric structures, inferred from observations and semi-empirical modelling, is given in Table 1. Since the temperature (T), density (of thermal electrons, n_e, and neutral hydrogen atoms, n_H) and magnetic field values (B) are derived from imaging observations, they are average quantities affected by the integration along and perpendicular to the line of sight. The magnetic field values cited for the corona are

Table 1. Typical averaged parameters of solar atmospheric structures

	T [K]	n_e [m^{-3}]	n_H [m^{-3}]	B[T]
Photosphere [a]				
Bottom	6500	$8 \cdot 10^{19}$	10^{23}	-
Sunspot	4000	$5 \cdot 10^{18}$	$4 \cdot 10^{23}$ [b]	0.3
Temp. min	4400	$3 \cdot 10^{17}$	$2 \cdot 10^{21}$	
Chromosphere				
Quiet[c]	10^4	$4 \cdot 10^{16}$	$5 \cdot 10^{16}$	$3 \cdot 10^{-3}$
Plage[d]	10^4	$3 \cdot 10^{17}$	$4 \cdot 10^{17}$	$3 \cdot 10^{-2}$
Corona				
Hole	$\leq 10^6$	10^{13}	-	10^{-4}
Active region	$4 \cdot 10^6$	10^{16}	-	10^{-2}
Quiet	$2 \cdot 10^6$	10^{14}	-	10^{-3}
Prominence	7000	10^{16}	10^{17}	10^{-3}

	$f_{\rm pe}$ [s^{-1}]	$f_{\rm ce}$ [s^{-1}]	$\nu_{\rm ep}$ [s^{-1}]	$\nu_{\rm en}$ [s^{-1}]	$\sigma[(\Omega{\rm m})^{-1}]$
Photosphere					
Bottom	$8 \cdot 10^{10}$	-	-	10^{10}	90
Sunspot	$2 \cdot 10^{10}$	$9 \cdot 10^9$	-	$3 \cdot 10^{10}$	2
Temp. min	$5 \cdot 10^9$		-	10^8	25
Chromosphere					
Quiet	$2 \cdot 10^9$	$9 \cdot 10^7$	10^6	$4 \cdot 10^3$	10^3
Plage	$5 \cdot 10^9$	$9 \cdot 10^8$	10^7	$3 \cdot 10^4$	10^3
Corona					
Hole	$3 \cdot 10^7$	$3 \cdot 10^6$	1	-	$7 \cdot 10^5$
Active region	$9 \cdot 10^8$	$3 \cdot 10^8$	10^2	-	$6 \cdot 10^6$
Quiet	$9 \cdot 10^7$	$3 \cdot 10^7$	3	-	$2 \cdot 10^6$
Prominence	$1 \cdot 10^9$	$3 \cdot 10^7$	10^6	$8 \cdot 10^3$	$8 \cdot 10^2$

[a] cf. Maltby et al. 1986, ApJ 306, 284
[b] Denser than the photospheric bottom because observed at a lower height
[c] cf. Vernazza et al. 1981, ApJS 45, 635
[d] cf. Lemaire et al. 1981, A&A 103, 160

consistent with present knowledge, but are in general not directly measurable (with the exception of prominences: see Sec. 5.2). f_{pe} and f_{ce} are the electron plasma frequency and the electron cyclotron frequency, respectively. Characteristic frequencies of momentum exchange are given for collisions of electrons with protons (ν_{ep}) and with neutral hydrogen (ν_{en}; after Melrose and Khan 1989). The indicated electric conductivity σ is the minimum value of those inferred from collisions of electrons with neutrals (photosphere; cf. Stix 1991, ch. 8.1.2) and with protons. Photospheric electrons are supplied by heavy atoms with low ionisation potential, while hydrogen is neutral. This is why no electron-proton collision frequency is listed for the photosphere.

2.3 Subphotospheric Motions as Drivers of Solar Activity

The evolution of the plasma structures and the measurement of photospheric magnetic fields show that the plasma β decreases with increasing height in the solar atmosphere, and increases again at heights of solar wind acceleration. The photosphere is a high-β plasma, where the pressure of the convected gas concentrates magnetic fields in individual flux tubes. Sunspots are the greatest and most conspicuous of them, but flux tubes of smaller spatial scale form a network in the photosphere on the borders of convective cells with a typical size of 30,000 km (supergranulation cells; $B \geq 0.1$ T). It seems that with increasing spatial resolution more concentrated magnetic fields become visible in the "quiet" photosphere.

Material motions observed in the photosphere reflect processes in the underlying convective zone. The density scale height in the convective zone below the photosphere goes from $\approx$ 100 Mm at the bottom to 0.2 Mm at the top. Compared to the height of the convective zone, this implies a density ratio between the bottom and top of $\approx 10^6$. This strong density stratification (coupled to mass conservation) creates a strong asymmetry between upward and downward motions: the ascending plasma must rapidly expand, then it must overturn within a density scale height while the descending plasma must rapidly contract and continue to fall down, becoming more concentrated and more dense. It implies that the convective zone has gentle ascending motions and strong concentrated downflows (just the opposite to the hot ascending plumes in the earth's upper mantle !). Convective motions concentrate magnetic fields in individual flux tubes, part of which may become buoyantly unstable and rise through the photosphere. The emergence of magnetic flux above the photosphere is visualized e.g. by "arch filament systems" (AFS). They are formed by several dark arches with a blue-shifted summit and red-shifted legs (e.g. Chou and Zirin 1988; Alissandrakis et al. 1990 Mein et al. 1996). The observed Doppler velocities are direct signatures of emerging magnetic loops with dense plasma leakage in the legs.

The coupling of the magnetic field in the atmosphere with material motions in and below the photosphere is the basic process driving the transport of energy into the corona. Motional electric fields for typical photospheric (horizontal) speeds of 0.5 km s^{-1} and magnetic fields of $\sim$100 G provide a Poynting flux of $5 \cdot 10^4$ Wm^{-2} (Einaudi and Velli 1994). The energy loss of the corona ranges from several 10^2 Wm^{-2} in a coronal hole to 10^4 Wm^{-2} in an active region. Therefore the Poynting flux is sufficient to explain coronal heating and solar wind acceleration. However, the power supplied to the active region corona (typical photospheric surface 10^{15} m^2) is 10^{20} W, which is not sufficient to power a conspicuous flare. A flare requires energy storage during up to a day, presumably in the magnetic field configuration.

3   Why an MHD Description of the Corona?

The convective zone governs the physics in the corona both by successive emergences of magnetic flux and by imposing horizontal photospheric motions. The coronal magnetic field and plasma are forced to evolve according to the time-dependent boundary conditions imposed at the photospheric level by the convective zone. This boundary driving is however slow at the bottom of the corona (few km.s^{-1}) compared to the sound and Alfvén speed (see Table 2) and it is fundamentally different from the super Alfvénic driving ($M_A \approx 10$) of the terrestrial magnetosphere by the solar wind. The sub-photospheric driving induces a quasistatic evolution of the coronal magnetic field which is interrupted from time to time by dynamical events. Part of the free-energy is either dissipated directly or stored for a short duration and released frequently in small events giving an "average" coronal heating. The other part of the free-energy is stored on a longer time (hours to days) and it is liberated only when a global instability of the magnetic configuration occurs; it leads to flares and CMEs.

A second major difference between the terrestrial magnetosphere and the solar corona is the plasma densities; they imply a collisionless plasma for the magnetosphere while in the corona the frequency of collisions is much higher than the bounce frequency (see Table 2). This permits to use MHD equations for analysing the large-scale magnetic configurations in the corona and their evolution. A third difference is the importance of the magnetic field in the physics involved: while the low corona is fully dominated by the magnetic field (low β plasma, see Table 2) particle pressure cannot always be neglected in the magnetosphere (see e.g. Fontaine 2000).

On top of these three main physical differences is the weight of the observing technics: localized measurements of particle distribution functions, fields and currents versus remote sensing imaging and spectrographic observations measuring density, velocity and magnetic field averaged over the instrumental response function. The MHD description of the corona attempts to understand the evolution of magnetic structures on the large spatial scales accessible to telescopes, and to infer where the macroscopic approach breaks down, i.e. where energy will eventually be released. The basic boundary conditions are magnetic fields and flow fields, B and v, measured at the photospheric

Table 2. Order of magnitude for the physical parameters in the low solar corona. The numerical values (third column) are given for a temperature $T = 10^6$ K, a density $N = 10^{15}$ m^{-3}, a magnetic field $B = 10^{-2}$ T (and $\ln \Lambda = 20$) which are typical for the low corona. Variation from these values can be easily computed by using the parameter dependence reported in the fourth column with the quantities T, N, B normalized to the above values. (From Chen 1984, NRL Memorandum Report 1977; Priest 1982; Stix 1991)

parameter	symbol	typical value	dependence
Lengths			
active region	L_{AR}	10^8 m	
scale height	H_g	$5 \cdot 10^7$ m	T
supergranule	L_{SG}	$3 \cdot 10^7$ m	
sunspot	L_{SS}	10^7 m	
granule	L_G	10^6 m	
Collision mean free path	$\lambda_{ee} \approx \lambda_{ep} \approx \lambda_{pp}$	$2 \cdot 10^5$ m	T^2/N
Plasma skin depth		$2 \cdot 10^{-1}$ m	$N_e^{-1/2}$
Proton gyro radius	r_p	10^{-1} m	$\sqrt{T_p}/B$
Debye length	λ_D	$2 \cdot 10^{-3}$ m	$\sqrt{T_e/N_e}$
Electron gyro radius	r_e	$2 \cdot 10^{-3}$ m	$\sqrt{T_e}/B$
Times			
Solar cycle		22 years	
active regions, filaments		10^7 s	
CMEs		10^5 s	
Flares: Main phase		10^4 s	
Impulsive phase		$3 \cdot 10^2$ s	
Hard X-rays spikes		10^{-1} s	
Radio spikes		10^{-2} s	
Frequencies			
Electron gyrofrequency	f_{ce}	$3 \cdot 10^8$ s^{-1}	B
Plasma frequency	f_{pe}	$3 \cdot 10^8$ s^{-1}	$\sqrt{N_e}$
Proton gyrofrequency	f_{cp}	$2 \cdot 10^5$ s^{-1}	B
Electron-electron collision rate	ν_{ee}	50 s^{-1}	$N_e T_e^{-3/2}$
Electron-proton collision rate	ν_{ep}	30 s^{-1}	$N_e T_e^{-3/2}$
Proton-proton collision rate	ν_{pp}	1 s^{-1}	$N_p T_p^{-3/2}$
Bounce frequency for electron	V_{Te}/L_{AR}	$4 \cdot 10^{-2}$ s^{-1}	$\sqrt{T_e}$
Bounce frequency for proton	V_{Tp}/L_{AR}	10^{-3} s^{-1}	$\sqrt{T_p}$

Table 2. Continued

parameter	symbol	typical value	dependance
Velocities			
particle beams		up to $3 \cdot 10^5$ km s^{-1}	
solar wind		≈ 700 km s^{-1}	
jets		few 100 km s^{-1}	
quiescent coronal evolution		$1 - 10$ km s^{-1}	
photospheric motions		$0.1 - 1$ km s^{-1}	
Electron thermal speed	V_{Te}	$4 \cdot 10^3$ km s^{-1}	$\sqrt{T_e}$
Alfven speed	V_A	10^3 km s^{-1}	$B/\sqrt{N}$
Sound speed	C_s	$2 \cdot 10^2$ km s^{-1}	$\sqrt{T}$
Proton thermal speed	V_{Tp}	10^2 km s^{-1}	$\sqrt{T_p}$
Plasma parameters			
Plasma parameter	$N_e \lambda_D^3$	10^7	$T_e^{3/2} N_e^{-1/2}$
Plasma beta	β	$4 \cdot 10^{-4}$	$N B^{-2} T$
Magnetic diffusivity	η	1 m^2 s^{-1}	$T^{-3/2}$

level. Magnetospheric diagnostics, on the other hand, provide local parameters from which the global configuration must be inferred. The difficulty to do this adequately with a kinetic approach is such that even in the magnetosphere MHD is frequently used in a global approach (e.g. Birn et al. 1996; Hesse et al. 1997). Obviously, in both fields kinetic theory, or even more a particle approach is needed when the detailed process of energy release is to be analyzed.

Even when the object of the study is the global understanding of the physical system, the "macroscopic" equations used, as well as the concepts behind, are different. The long standing controversy on whether one should mainly use MHD with B and v for the primary variables, or rather use E and j leads to many misunderstandings between the two communities. They are still well alive as the recent debate between Parker (1996, 1997) and Heikkila (1997) shows.

MHD equations express the conservation of fundamental macroscopic quantities (mass, impulsion and energy). They are derived from the Boltzmann equation by integrating over the velocity space, so there is some lost information (see e.g. Parker 1994; Heyvaerts 2000). In the study of solar atmospheric structures some simplifying approximations are well justified:

nonrelativistic regime, fluid approach (enough particles or dominant magnetic field), evolution time much longer than the cyclotron period of particles, absence of large-scale field-aligned potential drops. Some are more questionable, for example, an evolution time much larger than the thermalisation time of the particles. Moreover coefficients (like viscosity and resistivity) and the equation of state need to be specified in MHD equations. They are derived from kinetic theory in "ideal" cases (e.g. local thermal equilibrium) which are not necessarily valid in the applications. The relatively small mean free path of particles compared to global sizes (see Table 2) means that the Coulomb collisions are so frequent in the low corona that MHD may still be a good first approximation to study the processes involved.

One of the most troublesome parts of MHD may be the treatment of the relationship between current density and electric field. This is done through a generalised Ohm's law derived basically from the momentum equation of electrons. In the study of equilibrium configurations and their slow, quasi-static evolution in the coronal low β plasma an order of magnitude estimation of the terms leads usually to simplify this equation to retain only the resistive term ($j = \sigma E$, see e.g. Priest 1982). This is further justified by Parker (1994, Chapter 2) who showed that only some terms of the generalised Ohm's law have an effect on the magnetic field evolution and that such terms are usually negligible. In most analyses, including numerical simulations, the electrical conductivity reduces to a scalar. This is clearly an oversimplification which is imposed by the limited resources of present computers. Even using super-computers, the scalar conductivity used needs to be reduced by many orders of magnitude compared to the estimated coronal one if one wants to study a global magnetic configuration. This may look very far from coronal physics, but it is less than one might think. There are several studies on magnetic reconnection and MHD turbulence which show that the energy release depends only weakly (logarithmic dependance) on the magnitude of the dissipative coefficients (see e.g. Biskamp 1993). In fact, with lower dissipation coefficients, the energy simply cascades to smaller scales, where it is finally dissipated. The input of the energy at large scales is basically fixed by the global evolution of the system, e.g. a large scale instability.

There is an important restricted version of the MHD equations, so-called ideal MHD, which is free of the problems described above and can still give several hints on the magnetic-field physics. In ideal MHD both viscosity and resistivity are set to zero (or more precisely both Reynolds numbers are infinite). In the corona, one can even get a step further and neglect the plasma pressure (so no state equation is needed). This is a useful over-simplification because of the low β plasma and the high Reynolds numbers in the low corona. This simplified version of MHD is self-consistent (no need to define any plasma coefficient). This framework provides very valuable insight in how the whole magnetic system evolves. In particular it permits to understand how the energy is stored, when the system becomes globally unstable, so when the stored energy can be released. Moreover, it shows also where current sheets (or layers) are formed, so the spatial locations where magnetic energy can be efficiently transformed (by reconnection). The results can be directly tested with observations (Sect.4).

The ideal MHD approach shows also its own limits: the MHD equations are obviously invalid at the current layers. There, locally, a kinetic approach is required, but it needs to be coupled to the global configuration described by MHD, because both the spatial location and the intensity of the current layers are determined by the global evolution of the system. The resistive MHD can provide the global energy budget (see above), but cannot describe the energy release itself, for example the energization of particle beams. So for a full understanding of the processes both approaches, coupled together, are needed; this is a huge challenge ! At present there are still many unresolved questions on storage and sudden release of the magnetic energy, that MHD equations, even simplified ones, will still be of great help to understand the various solar observations.

4   Magnetic Energy Conversion in the Solar Atmosphere

The transformation of magnetic energy takes various forms in the solar atmosphere ranging from the de-stabilisation and ejection of a fraction of the corona down to the very small events implied in quasi-continuous coronal heating. Magnetic reconnection is thought generally to be at the heart of the energy conversion processes. In the following we emphasize the overall topology of the coronal magnetic field, as a basic concept to understand where energy could be released in a highly conductive medium. We do not discuss how energy is released. The relevant mechanisms involve sub-telescopic scales and non-MHD treatments, but they act in a macroscopic environment which can be realistically described by the MHD approach.

4.1   Magnetic Topology

Under typical coronal conditions the magnetic field is nearly force-free and frozen into the plasma almost everywhere in active regions on the Sun. An exception is separatrices which are magnetic surfaces where the magnetic field line linkage is discontinuous (see Fig. 5). A particularly important location for reconnection (in a classical view) is the intersection of two separatrices, called a separator. Most of the reconnection theories locate the energy release at or in the close vicinity of a separator, because a current sheet forms generally at a separator when the magnetic configuration evolves. The simplest example is a 2-D magnetic configuration with a neutral X-point (which is a particular case of separator) as in Fig. 5b. Adding a third perpendicular component

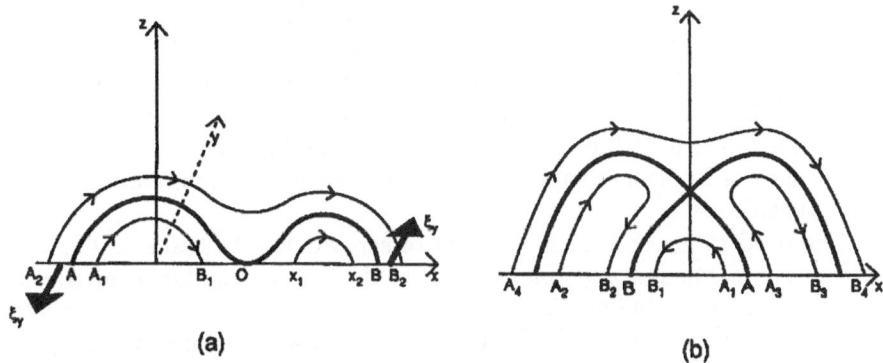

Fig. 5. Basic magnetic topology in a quadrupolar configuration. In (a) the separatrix (*thick line*) is tangent to the boundary at the point named "O" (it is a "bald patch"), while in (b) two separatrices intercept at an X point. In both cases the connectivity of field lines (*thin lines*) is discontinuous at the separatrices (as emphasis by the labeling of the foot-point of field lines). Shearing photospheric motions (*thick arrows*) induce the formation of a current sheet all along the separatrix in (a) (resp. separatrices in (b)). (From Vekstein and Priest 1992)

of the field, which is invariant in this perpendicular direction (so-called 2.5-D configurations), yields a new possibility for the current sheet formation. They can be formed now along the whole separatrices even when smooth shearing flows are present at the photospheric boundary (e.g. Zwingmann et al. 1985). This may occur in two distinct cases (see Fig. 5). Firstly when there is an X-point in the poloidal field (Low and Wolfson 1988; Finn and Lau 1991; Vekstein and Priest 1992). Secondly when there are field lines tangent to the photospheric boundary (Wolfson 1989; Low 1992; Vekstein et al., 1992). These locations of the photospheric inversion line where field lines are tangential and upward curved are called for brevity "bald patches" (BPs) by Titov et al. (1993).

In 3-D generic cases the separatrices generalize directly the above two classes. Separatrices are formed by field lines, which thread either null points or bald patches. Current sheets are thought to form along the separatrices when arbitrary motions are imposed at the photospheric plane (e.g. Aly 1990; Lau 1993). The generalization of 2-D null point to 3-D has received most of the attention because it defines a clear topology, both locally around one null and globally (the separator being the field line linking two nulls).

From observations, several flares have magnetic nulls in their reconstructed magnetic configuration but the relation of magnetic nulls with flares is not systematic (Démoulin et al. 1994). This was a first motivation for the generalization of separatrices to quasi-separatrix layers (QSLs). A second motivation, more theoretical, is summarized in the next paragraph. The notion of QSLs is related to some recent developments of 3-D magnetic reconnection theory. QSLs are the generalization of separatrices to magnetic configurations with a non-zero magnetic field strength everywhere in a region (Priest and Démoulin 1995). By definition the magnetic connectivity in QSLs enormously changes with slight variations of position of the field line footpoints. QSLs play a similar role as true separatrices, i.e. QSLs as well as separatrices are responsible for generating strong current layers and magnetic reconnection is expected even in plasma with a high magnetic Reynolds number (Démoulin et al. 1996a).

The need to generalize the concept of separatrices to QSLs is illustrated here with a particular example. Let's consider a quadrupolar magnetic configuration invariant in one direction, called y, like in Fig. 5b. The intersecting separatrices define four cells of connectivity. However, when the magnetic configuration has a finite extension in the y direction, there are no longer separatrices, in the cases with no bald patches and with a non-vanishing B_y component (so no magnetic null point). The structural instability of separatrices, when going from 2.5-D to 3-D, was first pointed out by Schindler et al. (1988) in the case of twisted magnetic configurations (with applications to the magnetospheric tail). However this structural instability is no longer present when the notion of separatrices is generalized to those of QSLs (Démoulin et al. 1996b).

4.2   Large Scale Magnetic Restructuring: Coronal Mass Ejections

Coronal Mass Ejections (CMEs) have been extensively observed in white light coronagraphs on Skylab, SOLWIND, SMM (Solar Maximum Mission) and presently with SOHO. Solar wind and CMEs are the two main hydromagnetic phenomena which eject plasma and magnetic fields out of the sun. A CME destabilizes a large part of the corona: its extension is usually of the order of one solar radius when it is first discernible in coronograph images (and its size in the interplanetary medium increases nearly linearly with distance from the sun). The ejected mass in one CME is typically of the order of $\approx 10^{12}$ kg, with a kinetic and gravitational energy of $\approx 10^{24} - 10^{25}$ J. The velocities are typically a few hundred km.s^{-1} up to one thousand km.s^{-1} (this upper value has the magnitude of the Alfvén velocity). While a flare is often observed in conjunction with a CME, it seems to start after the time when the backward extrapolation of the CME trajectory intersects the photosphere. The energy released in the flare is variable, but can be of the same order as the kinetic or potential energy of the CME. The relative timing of CMEs and flares shows that the mass ejection is not the consequence of the explosive energy release during the flare, but constitutes a specific manifestation of coronal magnetic field de-stabilisation.

A long standing question during the past thirty years dealt with the possibility to open a magnetic configuration. On one hand, the coronal plasma observations seem to indicate such opening, on the other hand the models have difficulties to explain such phenomena. The coronal magnetic Reynolds number being huge (10^{12}–10^{14}) on large scales, an ideal instability is more likely to drive the process initially. This problem was difficult to solve and it is only recently that a consensus has grown. A common result to all present numerical simulations is that the stored energy in a force-free field cannot exceed the energy of the associated open field (with the same photospheric magnetic-flux distribution; this provides a numerical validation to the conjecture of Aly (1991) and Sturrock (1991).

An important step in the analysis of the problem has been to consider axisymmetric magnetic configurations around a sphere rather than a cartesian geometry invariant by translation. The main difference is in the property of the open-field: it has a finite energy in the axisymmetric geometry while its energy is infinite in the cartesian geometry (even per unit of length in the invariant direction). Then photospheric shearing motions can drive the system to the open state in a finite time (compared to an infinite time in the cartesian 2.5-D geometry). This has been shown both by analytical (Lynden-Bell and Boily 1994; Aly 1995; Sturrock et al. 1995; Wolfson et al. 1996) and by numerical (Mikić and Linker 1994) approaches. The inclusion of resistivity allows the formation and ejection of a twisted flux tube. The actual precise amount of resistivity seems to play a minor role (see Sect. 3). The inclusion of solar wind makes the eruption more energetic (Linker and Mikić 1995). These works represent an important step compared to the cartesian case. In particular the field opens after a finite time, but this time, around 30 days with differential rotation, is too long compared to the observed rate of CMEs (about 1 per day).

Only recently has the analysis of the full 3-D problem become possible, due to the development of powerful numerical techniques and computers. Slow photospheric twisting motions applied to a simple bipolar field force the configuration to expand upward, initially at a speed lower than the driving speed, but later much faster (a non negligible fraction of the Alfvén speed; Amari et al. 1996). Because the twisting motions have been applied only to part of the photospheric field, the configuration tends to a partially open field in a finite time. The amount of twist needed, of about one turn, is more compatible with the observations than in the previous axisymmetric configurations. Finally, the full 3-D system has more freedom than the axisymmetric one: in particular the twisted flux tube can push the untwisted field lines aside, while in an axisymmetric system they would be forced to become open, too. This effect facilitates the partial opening of the field. This opening phenomenon has been found to occur in various configurations (Amari et al. 1997a).

4.3 Magnetic Field Evolution and Energy Dissipation During Flares

Flares are observed in the whole spectrum ranging form radio to γ-rays. There are a rich variety of magnetic configurations; models have to explain how and

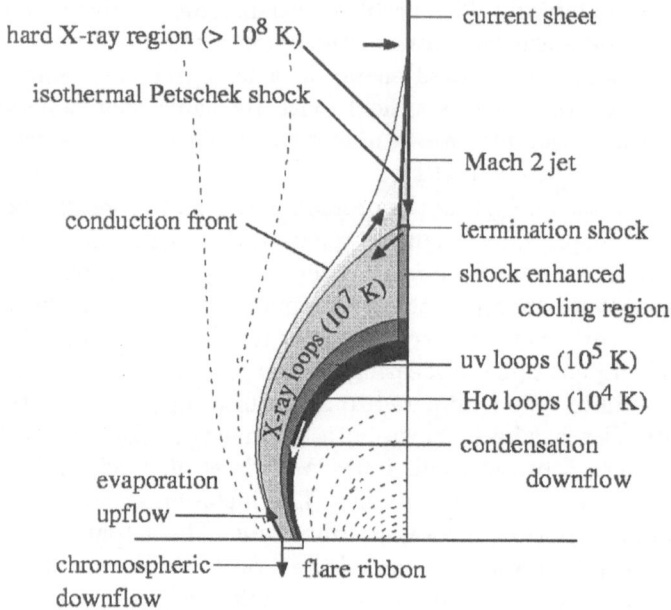

Fig. 6. Schematic diagram representing the characteristics of supermagnetosonic regime in flares together with the observational consequences. This regime is most likely to occur in the early phase of the flare when reconnecting fields are strong. Solid lines indicate boundaries between various plasma regions, while dashed lines indicate magnetic field lines. Arrows represent the plasma flows. (From Forbes and Acton 1996)

where magnetic energy is released in solar flares and in particular the existing link between chromospheric flare ribbons sometimes separated by more than 100 Mm. "Post" flare loops are seen to form between the flare ribbons during the development of most flares (see Schmieder 1992 for a review and Malherbe et al. 1997 for a time evolution of the phenomena) suggesting a model with reconnection (e.g. Forbes and Acton 1996; van Driel-Gesztelyi et al. 1997). Many observational studies in different wavelengths show that flares, and even less intense coronal phenomena, involve interactions between coronal magnetic structures (see e.g. Machado et al. 1988; Shimizu et al. 1994; Hanaoka 1995; van Driel-Gesztelyi et al. 1996). Figure 6 gives a summary of how the reconnection process is envisioned and how it may affect different regions of the solar atmosphere.

Since the instantaneous transport of energy from the convective region and the photosphere is not sufficient to power a flare, energy storage is a necessary ingredient of flare models. For example, it has been argued that current sheets can store enough magnetic energy to power a flare (Somov 1992 and references therein). At some point in the evolution, this current sheet

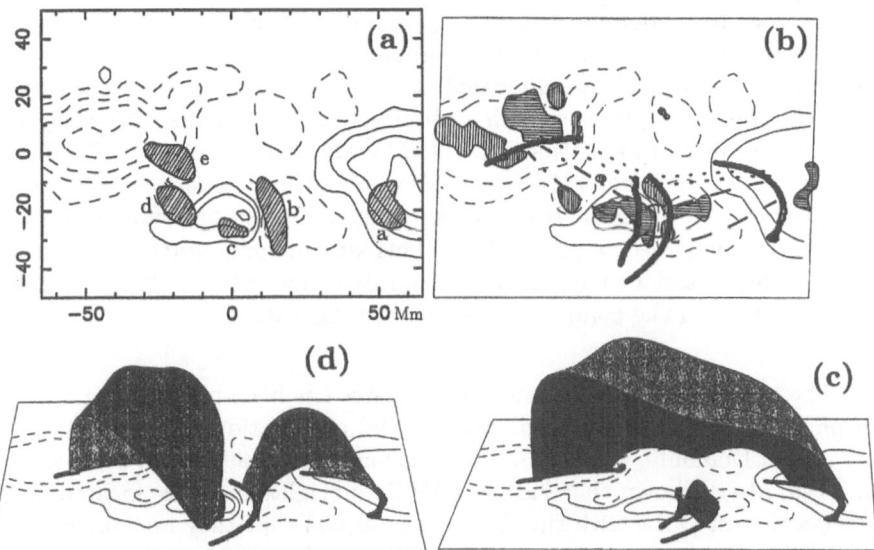

Fig. 7. Example of the correspondence between the flare ribbons and the trace of computed QSLs in a simple flaring region (a) Observational data: Off-band Hα flare kernels (hatched regions) and longitudinal photospheric field (positive and negative values are drawn with solid and dashed lines respectively) (b) Trace of the QSLs (thick lines) and regions where the vertical current density is greater than 10 mA m^{-2}. The coronal links between Hα brightenings are given by four kinds of field lines (the Hα kernel d is linked to local magnetic connections which are not represented). (c,d) Perspective view of Fig. 7b showing the coronal linkage at the borders of QSLs with field lines drawn as surfaces. (From Démoulin et al. 1997)

becomes unstable and turbulence develops increasing the plasma resistivity. Then, the stored energy is rapidly released as a flare (e.g. Heyvaerts et al. 1977). It has instead been argued that reconnection tends to occur at a rate imposed by the evolution of the large-scale magnetic field (e.g. Priest and Forbes 1992). In this latter case, the current in the sheet is always small and magnetic energy is instead stored in smooth field-aligned currents, such as a twisted flux tube, at a spatial scale that gives a negligible role to the resistive term (length-scale typically 1-10 Mm). In this evolution an ideal instability or non-equilibrium occurs forcing reconnection to take place at the separator (e.g. Priest and Forbes 1990). Another possibility is that the field-aligned currents are formed by photospheric or convective motions and then carried towards the locations of current sheets where the stored energy can be rapidly released. This list of possible solar flare models is far from being complete, but it shows instead that we still need to combine a large set of observations with adequate modeling of the magnetic field in a search about hints on the energy release process.

In order to model an observed region we need to compute the magnetic field from the photosphere to the corona. Important difficulties (linked to the presence of concentrated currents and of separatrices) are present in 3D force-free field computations using observed magnetograms. Currently they are still not fully solved (Amari et al. 1997b; Mc Clymont et al. 1997). Present successful extrapolations, in terms of a good correspondence between the observed coronal plasma structures and the computed field lines, have been realized for moderate magnetic shear. This kind of configuration provides the most accurate test of flare models in terms of localization of the magnetic energy release in the computed magnetic configuration.

The role of magnetic reconnection in flares has been tested using a variety of photospheric, chromospheric and coronal observations. By analysing several flares Démoulin et al. (1997), Mandrini et al. (1997), and Schmieder et al. (1997) have shown that Hα (or UV) flare brightenings are located on the intersection of QSLs with the chromosphere and that they are connected by magnetic field lines which trace the flare loops observed in soft X-rays. QSLs are formed in a variety of observed magnetic configurations, ranging from quadrupolar regions (an example is given in Fig. 7) to bipolar ones with an "S"-shaped inversion line and even in bipolar regions with a nearly potential field and an almost straight inversion line. The first case is a direct extension to 3-D of a 2-D magnetic configuration with an X-point (as used in 2-D reconnection models), while the second and third seem at first sight closer to a simple arcade model ! There is a wide range for the thickness of QSLs. This thickness is determined by the character (bipolar or quadrupolar) of the magnetic region and by the sizes of the photospheric field concentrations; the QSL thickness can be very small, ranging from 10^6 m down to zero (in the case of separatrices). Concentrated currents have been found in the observations; they are located at the borders of the QSLs (see Démoulin et al. 1997 and references therein). Moreover, two current kernels of opposite sign, linked by coronal field lines, are usually found at the photosphere. This indicates that the energy is presumably stored in the magnetic field associated with these field-aligned currents.

The above results give some confidence on the development of 3-D magnetic reconnection involving QSLs. They show that one needs to go beyond the classical generalization to 3-D of 2-D magnetic null points and associated separatrices. Magnetic reconnection occurs in more general circumstances when small scale lengths are formed in the system by a drastic change in the field line linkage. Magnetic configurations with field lines tangential to the photospheric boundary (at bald patches) can also lead to the formation of thin current layers (Sect.4.1). A first observational evidence of such a case has just been found (Aulanier et al. 1998c). It points to the need for the development of 3-D reconnection models in such a case.

4.4   Bright Points

X-ray bright points (XBPs) were first observed in images obtained by a rocket-borne grazing-incidence soft X-ray telescope (van Speybrock et al. 1970) and their characteristics were analysed in detail during the Skylab mission (e.g. Golub et al. 1977). They appear as diffuse clouds of typically 20 Mm diameter with a central bright core. They are uniformly distributed over the solar surface, with about 200 being present simultaneously and 1500 being born each day. Their lifetimes vary between 2 and 48 hours, with a mean value of 8 hours, and they are located above pairs of opposite magnetic polarities observed in the photosphere outside active regions (e.g. Harvey et al. 1994). Nolte et al. (1979) observed impulsive brightenings and rapid decays in a small sample of XBPs. They suggested that the fluctuations were driven by episodic heating superimposed upon a continuous input of energy. The cause of XBP variability is not yet clear, although some results indicate that the XBP brightenings are similar to normal flares, only smaller in energy (e.g. Strong et al. 1992), and may also involve beams of nonthermal electrons (Kundu et al. 1994).

Based on Skylab results, XBPs were originally considered the primary coronal manifestation of emerging photospheric flux. However, subsequent studies gave contradictory results concerning the nature of XBPs. Martin et al. (1985) suggested that cancelling features could be associated with XBPs. On the other hand, Golub et al. (1986) found that XBPs are more likely associated with emerging than with cancelling flux. More recently Webb et al. (1993), analyzing X-ray data from rocket flights coordinated with full-disk and time-lapse magnetograms, reached the conclusion that two-thirds of XBPs lie above decaying or cancelling magnetic features. Apart from the controversy, the important point to draw from these results is that the appearance of an XBP in the corona is independent of the type of feature seen in the magnetogram, as long as it is a bipole.

Parnell et al. (1994) showed that the analysed X-ray bright loops can be interpreted as being reconnected magnetic loops (computed from a simplified photospheric magnetogram). The process is driven by approaching magnetic flux-tubes of opposite polarity (see Parnell 2000 for further explanations). An MHD simulation of Dreher et al. (1997) confirms the formation of current sheets in such configurations and the subsequent magnetic reconnection. Mandrini et al. (1996) put forward the evidence of the role of magnetic reconnection in an XBP by using a direct extrapolation of the observed photospheric magnetic field and the QSL approach. The extrapolated field lines, with photospheric footpoints on both sides of QSLs, match the observed chromospheric and coronal structures (arch filament system, XBP and faint X-ray loops). Furthermore the calculated QSL is very thin (typically less than 100 m) during the lifetime of the XBP, but becomes much thicker ($\geq 10^4$ m) after the XBP has faded. This XBP shows an example of magnetic reconnection forced by emergence and displacement of a new bipole in an old magnetic field region, a situation which is also frequent in flares.

4.5   Coronal Heating

Energy is continuously supplied to the solar corona to maintain its temperature above 10^6 K. Since the major discovery of the existence of this very hot plasma in the 1940's, several mechanisms have been proposed. Nowadays a consensus has nearly emerged on the origin of the energy (which is thought to be injected at the photosphere as a Poynting flux) and on its mediator (namely the coronal magnetic field). The way this energy is dissipated in the corona is, however, still strongly debated (see e.g. Hollweg 1990; Gómez 1990; Einaudi and Velli 1994, for reviews). Two limiting cases have been investigated in detail: the case where the excitation time is comparable to or smaller than the Alfvén transit time of the coronal loops and the case where it is much larger. For the first case, a high-frequency excitation at the lower boundary, the dissipation of MHD waves by phase mixing (Heyvaerts and Priest 1983) or resonant absorption (Goossens 1991) is a prime candidate to heat the corona. For the second case, i.e. a low-frequency excitation, the formation and dissipation of thin current sheets has been proposed (Parker 1972). Because of the high magnetic Reynolds number of the corona, dissipation is very small on the typical scale lengths of the corona, and MHD turbulence is likely to be important both in the waves (e.g. Inverarity and Priest 1995) and quasi-static (e.g. Heyvaerts and Priest 1984) approaches. For example, granular and super-granular convective motions introduce perturbations on scales of ≈ 1 Mm and ≈ 30 Mm with time scales of $\approx 10^3$ and $\approx 10^5$ s, respectively. The dissipation of the perturbations on such time scales, with a classical resistivity, requires scale lengths $\approx 10^{-5}$ times smaller than the sizes of the convective cells for both types of convection (and so scale-lengths in the range 10–300 m). Clearly an efficient way to create very fine scale lengths is required.

The creation of current sheets naturally introduces fine scale-lengths in a magnetized plasma. Starting from a simple uniform field which is braided and twisted in an arbitrary way by photospheric motions, Parker (1972) argues that current sheets may be formed in ideal MHD conditions because there is, in general, no neighbouring equilibrium compatible with the imposed boundary conditions. This work stimulated a burst of research and controversy in the field (e.g. Aly 1987; Antiochos 1987; Zweibel and Li 1987; van Ballegooijen 1988; Longcope and Sudan 1992; Longcope and Strauss, 1994). Parker provided a formal example of the formation of a surface of tangential discontinuity when a layer of force-free field is locally compressed (Parker 1990), and presented a comprehensive development of the theory of the formation of discontinuities (Parker 1994).

Numerical experiments have demonstrated how quickly thin current concentrations tend to form. van Ballegooijen (1986) analysed the response of an initially uniform field to randomly phased boundary flows (modeling turbulent photospheric flows). Because of the non-linearity of the ideal MHD equations, the energy cascades to small scale lengths. van Ballegooijen (1988) showed that, even with simple boundary flows, the photospheric displacement of field-line footpoints contains fine structures with scale-lengths that become finer with time. These short scale-lengths are transferred to the coronal field and the electric-current density builds up in time. This has been confirmed by an ideal MHD simulation of Mikić et al. (1989) and by a resistive MHD simulation of Longcope and Sudan (1994) which show that current layers are formed with a thickness that decreases rapidly with time. Thin current layers are also formed during the dynamical evolution after the loss of equilibrium (Longcope and Sudan, 1992). Such non-equilibrium situations occur typically when the twist is between one and two turns (Longcope and Strauss 1994; Gómez et al. 1995).

X-ray observations show that the coronal plasma is highly inhomogeneous; this is present in the consequences of many models cited above. The spatial distribution of the heating is thought to be even more inhomogeneous than the present X-ray observations show (with a spatial resolution ≥ 0.7 Mm). Martens et al. (1985) for example have analysed a steady X-ray flaring loop-like structure and found a very hot component ($T \approx 10^7 K$); by modelling the energetics of the loops, they deduced that this component is present only in a tiny fraction ($\approx 10^{-3}$) of the volume and may be formed by about 30 current layers.

Curiously, while the heating is believed to be of magnetic origin, the regions with the highest magnetic field (above sunspots) are not bright in X-rays (e.g. Sams et al. 1992; Schmieder et al. 1996). Moreover, Metcalf et al. (1994) find no correlation between the locations of bright X-ray loops and the sites of strong photospheric currents! These observations tell us that neither the magnetic field nor the observable electric current are determining factors for the level of heating. The X-ray bright loops are rather observed around sunspots and above plages where the photospheric magnetic field is highly fragmented into thin flux tubes (see e.g. the review of Stenflo 1994). Démoulin and Priest (1997) show that such thin flux tubes imply the presence of a large number of very thin QSLs in the corona. A main parameter is the ratio between the magnetic flux located outside the flux tubes to the flux inside. The thickness of the QSLs is approximately given by the distance between neighbouring flux tubes multiplied by the ratio of fluxes to a power between two and three (depending on the density of flux tubes). They conclude that the fragmentation of the photospheric magnetic field stimulates the dissipation of magnetic energy in the corona.

Even restricting the photospheric excitation to low-frequency, there are at least five ways to form fine structures in the coronal magnetic field. Firstly, if the field topology is complex, there is current-sheet formation (see Sect.4.1). Secondly, fine structures can be introduced at the boundary because the link between the velocity pattern and the footpoint displacement is strongly non-linear: well-behaved flows produce fine structures with scale-lengths decreasing exponentially with time in the footpoint displacement (van Ballegooijen 1988). Thirdly, the continuous braiding by photospheric motions of coronal field can force the magnetic system to reach a non-equilibrium (e.g. Parker 1972; Berger, 1991; Longcope and Strauss 1994). Fourthly, the intrinsic non-linearity of the MHD equations introduces a cascade of energy to fine scales (e.g. Heyvaerts and Priest 1984; Dmitruk and Gómez 1997; Georgoulis et al. 1998). And, finally, the fragmentation in photospheric flux tubes introduces a very severe mapping distortion in the field-line linkage and creates many QSLs (Démoulin and Priest 1997). In conclusion we are far from a consensus on the mechanism of coronal heating ! Moreover, while there are several ways to form fine scale-lengths, the dissipation at these scales certainly requires more than the traditional MHD treatment. The recent discovery of different electron and ion temperatures provides a new diagnostic tool to study the relevant processes.

5   Prominences as Tracers of Coronal Magnetic Field Structures

5.1   Main Characteristics

Prominences (viewed at the limb as in Fig. 3), or filaments (viewed on the disk in Fig. 8a), are elongated structures of cold material (T $\approx$ 7000 K) suspended in the hot corona (T$\approx$ 10^6 K). They must be supported against gravity because their plasma is typically one hundred times denser than the coronal medium and because they extend in height over more than one hundred times the gravitational scale height of their cold plasma. Although most observations are carried out in lines of neutral atoms, the high collision rate between ions and neutrals effectively forces the neutrals to follow the ions (the relative velocity is only of a few m s^{-1} as shown by Mercier and Heyvaerts, 1977). Coronal magnetic fields are the usual explanation for prominence support. In fact the magnetic field has even a much broader impact in the physics of the prominences: it controls both the plasma dynamics (via the momentum equation) and thermodynamics (via the thermal conduction and the heating). This key role is not always recognized at its right level because the majority of observations focuses on the plasma. Furthermore, a direct evidence for importance of the magnetic-field is less compelling in prominences than in other phenomena like arch-filament systems, surges and coronal loops. Coronal loops consist basically of hot plasma which fills the magnetic flux tubes, while arch-filament systems and surges are dynamic plasma structures where dense plasma is forced to move along field lines. In both cases the plasma traces the magnetic flux tubes. At the opposite, during the quiescent phase of prominences the cold plasma fills only the extreme lower part of dipped

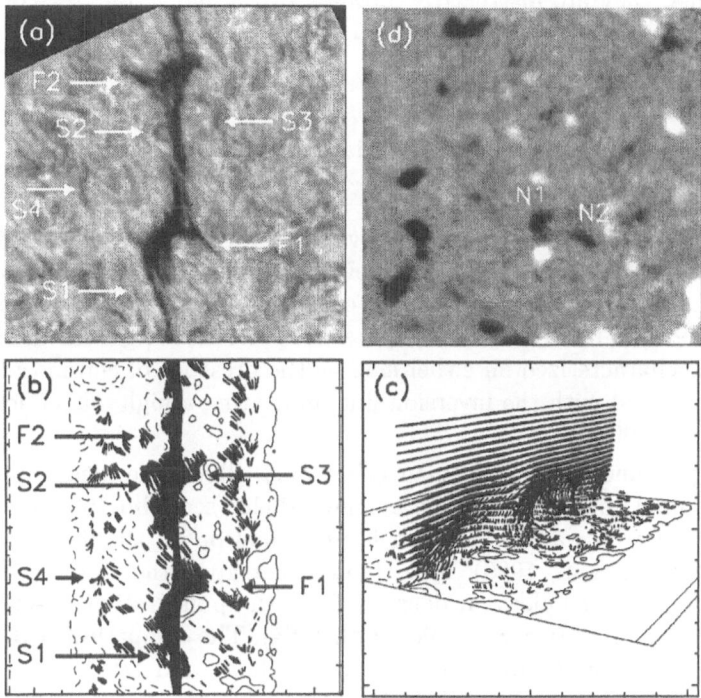

Fig. 8. (a) Example of a filament observed in the Hα line-center. Arrows point to particular structures (filament foot or chromospheric fibrils) (b) Model of the magnetic configuration computed from the photospheric magnetogram (represented by isocontours). The dark lines correspond to the lower bottom of the 3-D distribution of dipped field lines. (c) Side-view of the prominence model. The dark lines correspond only to the bottom of the dips which are supposed to be filled by dense plasma. Even in such a model, the twisted configuration is not apparent (see text). (d) Photospheric magnetogram from MDI/SOHO. (From Aulanier et al. 1998)

field lines (a view still under debate!); it is then difficult to relate directly the morphology of the cold plasma to the presence of the magnetic field. It is only when a detailed analysis of the various observations is realized in parallel with a theoretical approach that a coherent picture emerges.

This short description of filaments shows that they are very peculiar structures in the solar atmosphere both from the plasma and magnetic point of view. Still nowadays, the origin and the thermal stability of the cold plasma is a matter of research. Their magnetic structure is also still a matter of debate, partly because measurements of magnetic fields inside prominences yield counter-intuitive results (see below). There is in fact growing evidence that filaments are associated with the strongest non-potentiality of the coronal magnetic field. With the large set of observational techniques available in filaments, they are instructive "probes" of the more complex coronal structures.

5.2   Results from Magnetic-Field Measurements

Prominences are always found above lines where the vertical component of the photospheric magnetic field reverses sign. They are embedded within regions, called corridors or filament channels, which are nearly free of vertical magnetic field flux except small parasitic polarities (e.g. Martin 1990). They are also characterized on either side by the presence of chromospheric fibrils nearly aligned with the inversion line, indicating a high magnetic shear (e.g. Rompolt 1990).

In prominences the Zeeman effect only allows the measurement of the longitudinal component of the magnetic field (see Kim 1990 and references therein). Radio wavelengths provide information on the field strength (e.g. Apushkinskii et al. 1990). The Hanle effect gives the three components of the field (and the electron density) from the polarization measurements in two spectral lines (e.g. Bommier et al. 1994). The compatibility of the results obtained by these three independent methods and by different groups of observers have strongly contributed to validate the results (see Leroy 1988, 1989; Kim 1990).

One of the main results of Hanle measurements is that the prominence field has the opposite direction to the one expected from extrapolation of photospheric measurements (e.g. Leroy et al. 1983). Not only is the field component orthogonal to the prominence opposite to the field of a simple arcade (referred to as inverse configuration), but also the field component parallel to the prominence is opposite to those of an arcade that would have been sheared by differential rotation! This has been shown after a detailed analysis because twin solutions, symmetrical with regard to the line of sight, exist with optically thin lines and right angle scattering (known as 180^0 ambiguity). A large majority of prominences belong to the inverse type (75% in Leroy et al. 1984, 85% in Bommier et al. 1994 and greater than 90% in Bommier and Leroy 1998).

It is now well accepted that the magnetic field in prominences is nearly horizontal, while compatible with a slight magnetic dip (Bommier et al. 1994). The magnetic field strength is nearly homogeneous (Leroy 1989) on the scale of a few arc seconds, but shows a statistical increase of strength with height which is compatible with a large-scale dip configuration (e.g. Leroy et al. 1983).

5.3 Models for Prominence Support

The kind of magnetic configuration supporting filaments is still a matter of debate because the observations of the magnetic field are only partial (at the photosphere and in the prominence). Because filaments are long lived structures (weeks to months) there is clearly the need for a stable support. The most plausible one is the presence of a magnetic dip where dense plasma can be caught to form a filament (Kippenhahn and Schlüter 1957). There are three basic configurations which satisfy this constraint.

In arcade-like magnetic configurations a dip cannot be present in a force-free 2.5-D arcade, since the field lines become only flatter as the magnetic shear increases (Amari et al. 1991). This is however possible in 3-D with an overlying arcade compressing locally the central-part of an underlying sheared arcade (Antiochos et al. 1994). The latter gives mostly an inverse-polarity prominence with a magnetic field nearly aligned with the photospheric inversion line.

The second possibility is a support in quadrupolar configurations. Kippenhahn and Schlüter (1957, in their Sect. 4) first proposed these inverse configurations for stable support of dense plasma. The model was further developed by Malherbe and Priest (1983), Démoulin and Priest (1993), Drake et al. (1993), and Uchida (1998). The presence of a corridor free of significant field is needed to have a prominence extension reaching the chromosphere and converging motions are required to provide mass supply. The quadrupolar model has been extended to magnetic configurations typically found in active regions (Titov et al. 1993; Bungey et al. 1996) and in polar crown regions (Cartledge et al. 1996).

The third possibility, and the most plausible in view of the various observational constraints is the presence of a twisted magnetic configuration. It can be formed in several ways: by photospheric twisting motions (e.g. Priest et al. 1989), by converging motions in a sheared arcade with magnetic reconnection at the inversion line (e.g. van Ballegooijen and Martens 1989), by resistive instability in a sheared arcade (e.g. Inhester et al. 1992), by relaxation and accumulation of magnetic helicity (e.g. Rust and Kumar 1994) or by emergence from the convective zone (e.g. Low 1996). Low and Hundhausen (1995) show how a twisted-flux tube topology can bring together many of the chromospheric and magnetic observations. This was further developed for 3-D configurations by Aulanier and Démoulin (1998a). In particular they could reproduce naturally the feet of prominences which have been a long-standing puzzle (Fig. 8). This model has been successfully tested on an observed filament by computing the magnetic configuration associated to the photospheric magnetograms (Aulanier et al. 1998b).

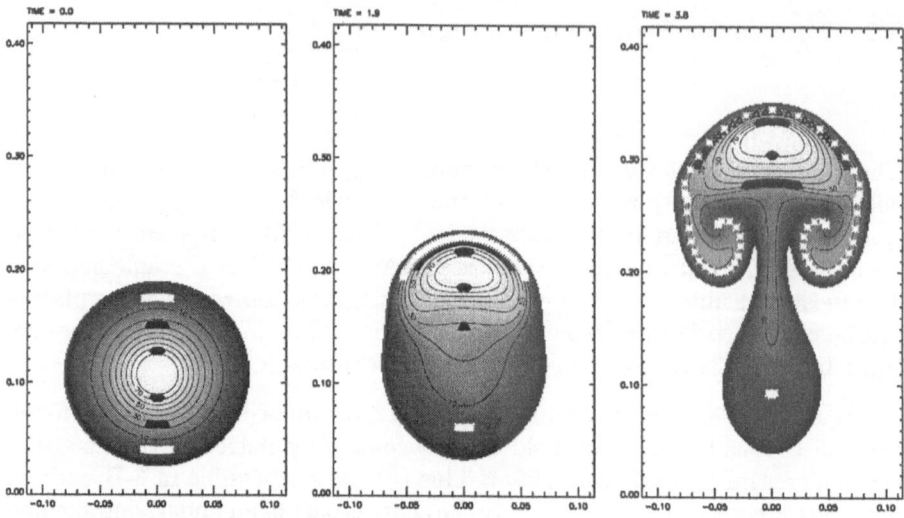

Fig. 9. Rise of a twisted flux tube in the convective zone with an initial magnetic field inclined to the flux tube axis by 7^0. The initial circular flux tube is deformed by the wake. Continuous lines represent isocontours of the field component along the flux tube axis. White and black tracers show the motions of individual plasma elements. (From Emonet and Moreno-Insertis 1998)

6   Global Evolution of Twisted Magnetic-Flux Tubes

6.1   Evolution in the Convective Zone

The amplification of magnetic field by plasma motions (dynamo mechanism) is taking place mainly at the bottom of the convective zone in the convective overshoot region (with a height estimated to ≈ 10 Mm). There the subadiabatically stratified plasma provides both plasma motions (in particular differential rotation) and mechanical stability for an intensification of the magnetic field (e.g. Spiegel and Weiss 1980). A Rayleigh-Taylor instability permits to form flux tubes. Above a strength ≈ 10 T the flux tube becomes undulatory unstable: the plasma in upward displaced parts moves to the downward displaced parts, leading to a buoyancy force acting upward on the upper parts of the flux tube (e.g. Spruit and van Ballegooijen 1982). Emonet and Moreno-Insertis (1998 and references therein) have shown that a minimum critical twist is needed so that the buoyant flux tube is not destroyed in its rise by the hydrodynamic vortex which develops behind. Moreover, for twist higher than the critical one, the combination of a higher buoyancy force in the central part of the flux tube with the effect of the following roles (in the wake) deform the initial twisted flux tube: it forms a magnetic configuration which has dips not only below the central O point but also on both sides (Fig. 9). These results have been confirmed by Fan et al. (1998).

6.2   Emergence at the Photosphere

So far the MHD simulations studied the deep convection zone and not the emergence through the photosphere of the flux tube. There the hypothesis breaks down: in particular the flux-tube radius becomes larger than the gravitational scale height. One possibility is that the flux tube splits in several smaller flux tubes. It is also not yet fully clear how the mass unloading of the flux tube is realized through this emergence but it is thought to be a difficult and long process (see e.g. Low 1996). The emergence of twisted configurations is supported by recent vector field measurements (Lites et al. 1995; Leka et al. 1996).

6.3   Evolution in the Corona

In the soft X-ray range the Yohkoh satellite provides many examples of S-shaped coronal loops which can be interpreted as the eruption of a twisted structure (see, e.g., Manoharan et al. 1996; Pevtsov et al. 1996; Rust and Kumar 1996). This eruption can be triggered by a loss of equilibrium or a rapid injection of new flux. In the last case, the eruption is driven directly from the emergence without need of pre-stored energy in the corona (Chen 1996), while in the first case, a slow photospheric evolution permits accumulation of energy in the corona long before the loss of equilibrium occurs. A catastrophe may happen in the configuration when a cusp in the equilibrium curve (e.g., twist versus height) is present (e.g. Lin et al. 1998). Finally the eruption of a 3-D twisted magnetic configuration on the sun may have close analogy with flux rope formation occurring in the Earth's magnetotail (e.g. Birn and Hesse 1990)

Many observations suggest a helical-like pattern during eruption of prominences: on the disk (e.g. Raadu et al. 1988) and more often at the limb (e.g. Rompolt 1990; Vršnak et al. 1991). During the quiescent phase, prominences and filaments show little direct evidence of the general magnetic configuration supporting them (see Sec. 5.1). However a model based on the photospheric magnetogram can permit to recover the shape of the prominence with a twisted flux tube (Aulanier et al. 1998b). Moreover the magnetic topology found is basically the one found independently in MHD simulations of Emonet and Moreno-Insertis (1998, see in Sec. 6.1) !

6.4   Ejection in the Interplanetary Space

Observation of CMEs can be frequently associated to the eruption of a prominence. This relationship is confirmed both by the detailed examination of well-observed events and by statistical analysis (e.g. Hundhausen 1988). In coronograph pictures the prominence embedded within the coronal CME material often displays a distinctly twisted structure which strongly suggests a flux rope structure of the expanding magnetic field. Other but morphological evidence for this comes e.g. from radio observations. In a limb event Klein and Mouradian (1991) observed that the structure initially seen as a rising prominence shows later up as a radio source where confined electrons rise, pursuing the trajectory of the rising prominence in a uniformly accelerated ascending motion that extends over several solar radii. The prolonged confinement of the electrons suggests the presence of flux rope type twisted field lines in the rising structure. In some particularly favorable observations the ejection can be followed to large distances: with different instruments, Jackson et al. (1988) have followed a prominence eruption and the associated CME from the solar surface up to 100 $R_\odot$!

The twisted magnetic flux tube is also in agreement with the topology inferred for CMEs from coronagraph observation(e.g. Hundhausen 1988; Chen et al. 1997; Simnett et al. 1997). In the interplanetary medium, twisted configurations are also identified in magnetic clouds (or interplanetary CMEs) with in situ measurements from Ulysses (e.g. Bothmer et al. 1996; Weiss et al. 1996; Farrugia 1997; Osherovich and Burlaga 1997). The link between magnetic clouds, CMEs and prominence eruptions is highly probable both on a statistical ground (e.g. Bothmer and Schwenn 1994; Marubashi 1997) and from the detailed study of individual cases (e.g. Burlaga et al. 1998).

In conclusion, observational evidence and theoretical investigations from the convective zone to the interplanetary space are in favor of the formation and subsequent ejection of twisted flux tubes (After all they are natural configurations for a stressed magnetic field !). The twisted flux tubes, probably formed at the bottom of the convective zone, bring both magnetic energy and helicity in the corona. In a highly conductive medium like the corona, magnetic energy can be dissipated at a fast rate (which is only weakly dependent on the magnetic Reynolds number), while magnetic helicity is a well preserved quantity (see Biskamp 1993). In this way energy release, which is confined in the corona, permits to get rid of part of the magnetic energy excess but cannot remove the magnetic helicity (and the associated excess of magnetic energy). Most of this helicity cannot be removed by cancellation of opposite helicities because the sign of helicity is mainly hemisphere dependent (negative/positive in the north/south hemisphere) without the reversal of sign observed for the magnetic field itself after each solar cycle (11 years). The only way the sun has to get rid of the accumulated magnetic helicity is then by ejecting it in the interplanetary medium (Low 1996). Such evolution (from the formation to the ejection) of twisted flux tubes is a natural, though complex, consequence of the MHD equations.

References

1. Alissandrakis, C.E. (1994). In: Belvedere, G., Rodonó, M., Simnett, G.M. (Eds.) Advances in Solar Physics, Lecture Notes in Physics, Springer, Berlin, **432**, 109
2. Alissandrakis, C.E., Tsiropoula, G., Mein, P. (1990) A&A **230**, 200

3. Aly, J.J. (1987). In: Beck, R., Gräve, R. (Eds.) Interstellar Magnetic Fields, Springer Verlag, Berlin, 240
4. Aly, J.J. (1990). In: Dezsö, L. (Ed.) The Dynamic Sun, Public. Debrecen Obs., 176
5. Aly, J.J. (1991) ApJ **375**, L61
6. Aly, J.J. (1995) ApJ **439**, L63
7. Amari, T., Démoulin, P., Browning, P., et al. (1991) A&A **241**, 604
8. Amari, T., Luciani, J.F., Aly, J.J., Tagger, M. (1996) ApJ **446**, L39
9. Amari, T., Luciani, J.F., Aly, J.J., Mikić, Z. (1997a). In: Crooker, N. Joselyn, J.A., Feynman, J. (Eds.) Coronal mass ejections, American Geophysical Union, 101
10. Amari, T., Aly, J.J., Luciani, J.F., Boulmezaoud, T., Mikić, Z. (1997b) Solar Phys. **174**, 129
11. Antiochos, S.K. (1987) ApJ **312**, 886
12. Antiochos, S.K., Dahlburg, R.B., Klimchuck, J.A. (1994) ApJ **420**, L41
13. Apushkinskii, G.P., Nesterov, N.S., Topchilo, N.A., Tsyganov, A.N. (1990), SvA, **34(5)**, 530
14. Aulanier, G., Démoulin, P. (1998a) A&A **329**, 1125
15. Aulanier, G., Démoulin, P., van Driel-Gesztelyi, L., et al. (1998b) A&A **335**, 309
16. Aulanier, G., Démoulin, P., Schmieder, B., et al. (1998c) Solar Phys. **183**, 369
17. Berger, M.A. (1991) A&A **252**, 369
18. Benz, A.O. (1993) Plasma Astrophysics, Kluwer
19. Bird, M.K., Edenhofer, P. (1990). In: Harve Schwenn, R., Marsch, E. (Eds.) Physics of the Inner Heliosphere Vol. 1, Springer, Berlin, Heidelberg, New York, 13
20. Birn, J., Hesse, M. (1990) Geophys. Monogr. Ser., **58**, 655
21. Birn, J., Hesse, M., Schindler, K. (1996) JGR **101, A6**, 12939
22. Biskamp, D. (1993) Non linear MHD, Cambridge University Press, Cambridge
23. Bommier, V., Leroy, J.L. (1998). In: Webb, D., Schmieder, B., Rust, D. (Eds.) IAU Colloq. 167, Astronomical Society of the Pacific, **150**, 434
24. Bommier, V., Landi Degl'Innocenti, E., Leroy, J.L., Sahal-Bréchot S. (1994) Solar Phys. **154**, 231
25. Bothmer, V., Schwenn, R. (1994) Spa. Sci. Rev., **70**, 215
26. Bothmer, V., Desai, M.I., Marsden, R.G., et al. (1996) A&A **316**, 493
27. Bungey, T.N., Titov, V.S., Priest, E.R., (1996) A&A **308**, 233
28. Burlaga, L., Fitzenreiter, Lepping, R., et al. (1998) JGR **103, A1**, 277
29. Cartledge, N.P., Titov, V.S., Priest, E.R. (1996) Solar Phys. **166**, 287
30. Chen, J. (1996) JGR **101**, 27499
31. Chen, J., Howard, R.A., Brueckner, G.E., et al. (1997) ApJ **490**, L191
32. Chou, D.Y., D.Y., Zirin, H. (1988) ApJ **333**, 420
33. Cox, A.N., Livingston, W.C., Matthews, M.S. (Eds.), (1991) Solar Interior and Atmosphere, The Univ. Arizona Press, Tucson
34. Crooker, N., Joselyn, J.A., Feynman, J. (Eds.), (1997) Coronal mass ejections, Geophysical Monograph Series
35. David, C., Gabriel, A.H., Bely-Dubau, F., et al. (1998) A&A **336**, L90
36. Démoulin, P., Priest, E.R. (1993) Solar Phys. **144**, 283
37. Démoulin, P., Priest, E.R. (1997) Solar Phys. **175**, 123
38. Démoulin, P., Hénoux, J.C., Mandrini, C.H. (1994) A&A **285**, 1023

39. Démoulin, P., Hénoux, J.C., Priest, E.R., Mandrini, C.H. (1996a) A&A **308**, 643
40. Démoulin, P., Priest, E.R., Lonie, D.P. (1996b) JGR **101**, A4, 7631
41. Démoulin, P., Bagalá, L.G., Mandrini, C.H., et al. (1997) A&A **325**, 305
42. Dmitruk, P., Gómez, D.O. (1997) ApJ **484**, L83
43. Drake, J.F., Mok, Y., van Hoven, G. (1993) ApJ **413**, 416
44. Dreher, J., Birk, G.T., Neukirch, T. (1997) A&A **323**, 593
45. Einaudi, G., Velli M. (1994). In: Belvedere, G., Rodonó, M., Simnett, G.M. (Eds.) Advances in Solar Physics, Lecture Notes in Physics, Springer, Berlin, **432**, 149
46. Emonet, T., Moreno-Insertis, F. (1998) ApJ **492**, 804
47. Fan, Y., Zweibel, E.G., Lantz, S.R. (1998) ApJ **493**, 480
48. Farrugia, C.J. (1997). In: Crooker, N. Joselyn, J.A., Feynman, J. (Eds.) Coronal mass ejections, American Geophysical Union, 177
49. Finn, J.M., Lau, Y.T. (1991) Phys. Fluids **B3**, 2675
50. Fontaine, D. (2000) this volume
51. Forbes, T.G., Acton, L.W. (1996) ApJ **459**, 330
52. Georgoulis, M.K., Velli, M., Einaudi, G. (1998) ApJ **497**, 957
53. Golub, L., Pasachoff, Jay M. (1997) The Solar Corona, Cambridge University Press
54. Golub, L., Harvey, J.W., Webb, D.F. (1986). In: Poland, A.I. (Ed.), Coronal and Prominence Plasmas, NASA CP-2442, 365
55. Golub, L., Krieger, A.S., Harvey, J.W., Vaiana, G.S. (1977) Solar Phys. **53**, 111
56. Goossens, M. (1991). In: Priest, E.R., Hood, A. W. (Eds.) , Advances in Solar System Magnetohydrodynamics, Cambridge University Press, Cambridge 137
57. Gómez, D.O. (1990) Fund. Cosmic Phys., 14, 131
58. Gómez, D.O., DeLuca, E.E., Mc Clymont, A.N. (1995) ApJ **448**, 954
59. Hanaoka, Y. (1995) Solar Phys. **165**, 275
60. Harvey, K.L., Nitta, N., Strong, K.T., Tsuneta, S. (1994). In: Uchida, Y., Watanabe, T. Shibata, K., Hudson, H.S. (Eds.) X-ray Solar Physics from Yohkoh, Frontiers Science Series, 10, 21
61. Heikkila, W.J. (1997) JGR **102, A5**, 9651
62. Heyvaerts, J. (2000) this volume
63. Heyvaerts, J., Priest, E.R. (1983) A&A **117**, 220
64. Heyvaerts, J., Priest, E.R. (1984) A&A **137**, 63
65. Heyvaerts J., Priest E.R., Rust D.M. (1977) ApJ **216**, 123
66. Hesse, M., Birn, J., Hoffman, R.A. (1997) JGR **102, A5**, 9543
67. Hollweg, J.V. (1990) Comput. Phys. Rep., **12**, 205
68. Hundhausen, A.J. (1988). In: Pizzo, V.J., Holzer, T.E., Sime, D.G. (Eds.) Solar Wind 6, NCAR TN-306, Boulder **1**, 181
69. Inhester, B., Birn, J., Hesse, M. (1992) Solar Phys. **138**, 257
70. Inverarity, G.W., Priest, E.R. (1995) A&A **302**, 596
71. Jackson, B., Rompolt, B., Svestka, Z. (1988) Solar Phys. **115**, 327
72. Kim, I.S. (1990). In: Ruždjak, V., Tandberg-Hanssen, E. (Eds.) IAU Colloq. 117, Springer, Berlin, 49
73. Kippenhahn, R., Schlüter, A. (1957) Zs. Ap. **43**, 36
74. Klein, K.-L. (1992). In: Méthodes de détermination des champs magnétiques solaires et stellaires, Faurobert–Scholl, M., Frisch, H., Mein, N. (Eds.) Publ. Obs. Paris, 113

75. Klein, K.-L., Mouradian, Z. (1991). In: Schmieder, B., Priest, E.R. (Eds.) Proc. Chantilly Meeting on Solar Flares, Publ. Obs. Paris, 185
76. Koutchmy, S. (1994) Adv. Space Res. **14(4)**, 29
77. Kundu, M.R., Strong, K.T., Pick, M., et al. (1994) ApJ **427**, L59
78. Lau, Y.T. (1993) Solar Phys. **148**, 301
79. Leka, K.D., Canfield, R.C., McClymont, A.N., van Driel-Gesztelyi, L. (1996) ApJ **462**, 547
80. Leroy, J.L. (1988). In: Altrock, R.C. (Ed.) Solar and Stellar coronal structures and dynamics, National Solar Observatory, 422
81. Leroy, J.L. (1989). In: Priest, E.R. (Ed.) Dynamics and Structure of Quiescent Solar Prominences, Kluwer, Dordrecht, 77
82. Leroy, J.L., Bommier, V., Sahal-Bréchot, S. (1983) Solar Phys. **83**, 135
83. Leroy, J.L., Bommier, V., Sahal-Bréchot, S. (1984) A&A **131**, 33
84. Lin, J., Forbes, T.G., Isenberg, P.A., Démoulin, P. (1998) ApJ **417**, 368
85. Linker, J.A., Mikić, Z. (1995) ApJ **438**, L45
86. Lites, B.W., Low, B.C., Martinez Pillet, V., et al. (1995) ApJ **446**, 877
87. Longcope, D.W., Strauss, H.R. (1994) ApJ **437**, 851
88. Longcope, D.W., Sudan, R.N. (1992) ApJ **384**, 305
89. Low, B.C. (1992) A&A **253**, 311
90. Low, B.C. (1996) Solar Phys. **167**, 217
91. Low, B.C., Hundhausen, J.R. (1995) ApJ **443**, 818
92. Low, B.C., Wolfson, R. (1988) ApJ **324**, 574
93. Lynden-Bell, D., Boily, C. (1994) Mon. Not. R. Astron. Soc., **267**, 146
94. Machado, M.E., Moore, R.L., Hernández, A.M., et al. (1988) ApJ **326**, 425
95. Malherbe, J.M., Priest, E.R. (1983) A&A **123**, 80
96. Malherbe, J.M., Tarbell, T., Wiik, J.E., et al. (1997) ApJ **482**, 535
97. Mandrini, C.H., Démoulin, P., van Driel-Gesztelyi, L., et al. (1996) Solar Phys. **168**, 115
98. Mandrini, C.H., Démoulin, P., Bagalá, L.G., et al. (1997) Solar Phys. **174**, 229
99. Manoharan, P.K., van Driel-Gesztelyi, L., Pick, M., Démoulin, P. (1996) ApJ **468**, L73
100. Martin, S.F. (1990). In: Ružjdak, V., Tandberg-Hanssen, E. (Eds.) IAU Colloq. 117, Springer-Verlag, Berlin, 1
101. Martin, S.F., Livi, S.H.B., Wang, J., Shi, Z. (1985). In: Hagyard, M.J. (Ed.) Measurements of Solar Vector Magnetic Fields, NASA CP-2374, 403
102. Martens, P.C.H., Van den Oord, G.H.J., Hoyng P. (1985) Solar Phys. **96**, 253
103. Marubashi, K. (1997). In: Crooker, N. Joselyn, J.A., Feynman, J. (Eds.) Coronal mass ejections, American Geophysical Union, 147
104. Mc Clymont, A.N., Jiao, L., Mikić, Z. (1997) Solar Phys. **174**, 191
105. Mein, P., Démoulin, P., Mein, N., et al. (1996) A&A **305**, 343
106. Melrose D.B., Khan J.I. (1989) A&A **219**, 308
107. Mercier, C., Heyvaerts, J. (1977) A&A **61**, 685
108. Metcalf, T.R., Canfield, R.C., Hudson, H.S., et al. (1994) ApJ **428**, 860
109. Mikić, Z., Schnack, D.D., Van Hoven, G. (1989) ApJ **338**, 1148
110. Mikić, Z., Linker, J.A. (1994) ApJ **430**, 898
111. Moore, R.L., La Rosa, T.N., Orwig, L.E. (1995) ApJ **438**, 985
112. Nolte, J.T., Solodyna, C.V., Gerassimenko, M. (1979) Solar Phys. **63**, 113
113. Osherovich, V., Burlaga, L.F. (1997). In: Crooker, N. Joselyn, J.A., Feynman, J. (Eds.) Coronal mass ejections, American Geophysical Union, 157

114. Parker, E.N. (1972) ApJ **174**, 499
115. Parker, E.N. (1990) Geophys. Astrophys. Fluid Dyn. **52**, 183
116. Parker, E.N. (1994) Spontaneous Current Sheets in Magnetic Fields, Oxford University Press, Oxford
117. Parker, E.N. (1996) JGR **101, A5**, 10587
118. Parker, E.N. (1997) JGR **102, A5**, 9657
119. Parnell, C.E. (2000) this volume
120. Parnell, C.E., Priest, E.R., Golub, L. (1994) Solar Phys. **151**, 57
121. Pevtsov, A.A., Canfield, R.C., Zirin, H. (1996) ApJ **473**, 533
122. Priest, E.R. (1982) Solar Magneto-hydrodynamics, D. Reidel, Dordrecht
123. Priest, E.R., Démoulin, P. (1995) JGR **100**, A12, 23443
124. Priest, E.R., Forbes, T.G. (1990) Solar Phys. **126**, 319
125. Priest, E.R., Forbes, T.G. (1992) JGR **97**, A11, 16757
126. Priest, E.R., Hood, A.W., Anzer, U. (1989) ApJ **344**, 1010
127. Raadu, M.A., Schmieder, B., Mein, N., Gesztelyi, L. (1988) A&A **197**, 289
128. Rompolt, B. (1990), Hvar Obs. Bull. Vol. **14, 1**, 37
129. Rust, D.M., Kumar, A. (1994) Solar Phys. **155**, 69
130. Rust, D.M., Kumar, A. (1996) ApJ **464**, L199
131. Sams, III B.J., Golub, L., Weiss, N.O. (1992) ApJ **399**, 313
132. Schindler, K., Hesse, M., Birn, J. (1988) JGR **93**, 5547
133. Schmelz, J.T., Brown, J.C. (1992) The Sun - A Laboratory for Astrophysics, NATO ASI Series, Kluwer
134. Schmieder, B. (1992). In: Svestka, Z., Jackson, B.V., Machado, M.E., (Eds.) Eruptive solar flares, Lecture Notes in Physics, Springer, Berlin. 124
135. Schmieder, B., Démoulin P., Aulanier G., Golub L. (1996) ApJ **467**, 881
136. Schmieder, B., Aulanier, G., Démoulin, P., et al. (1997) A&A **325**, 1213
137. Schwenn, R. (1990). In: Schwenn, R., Marsch, E. (Eds.) Physics of the Inner Heliosphere Vol. 1, Springer, Berlin, Heidelberg, New York, 99
138. Shimizu, T., Tsuneta, S., Acton, L.W., et al. (1994) ApJ **422**, 906
139. Simnett, G.M., Tappin, S.J., Plunkett, S.P., et al. (1997) Solar Phys. **175**, 685
140. Somov, B.V. (1992) Solar Flares, Springer, Berlin
141. Spiegel, E.A., Weiss, N.O. (1980) Nat **287**, 616
142. Spruit, H.C., van Ballegooijen A.A. (1982) A&A **106**, 58
143. Stenflo, J.O. (1994) Solar Magnetic Fields – Polarized Radiation Diagnostics, Kluwer, Dordrecht
144. Stix, M. (1991) The Sun, An Introduction, Springer, Berlin
145. Strong, K.T., Harvey, K.L., Hirayama, T., et al. (1992) Publ. Astron. Soc. Japan, **44**, L161
146. Sturrock, P.A. (1991) ApJ **380**, 655
147. Sturrock, P.A., Antiochos S.K., Roumeliotis G. (1995) ApJ **443**, 804
148. Tandberg-Hanssen, E., Emslie, A.G. (1988) The Physics of Solar Flares, Cambridge University Press, Cambridge, UK
149. Tandberg-Hanssen, E. (1995) The Nature of Solar Prominences, Kluwer, Dordrecht
150. Titov, S., Priest, E.R., Démoulin, P. (1993) A&A **276**, 564
151. Tu, C.-Y., Marsch, E., Wilhelm, K., Curdt, W. (1988) ApJ **503**, 475
152. Uchida, Y. (1998). In: Webb, D., Schmieder, B., Rust, D. (Eds.) IAU Colloq. 167, Astronomical Society of the Pacific, **150**, 163

153. van Ballegooijen, A.A. (1986) ApJ **311**, 1001
154. van Ballegooijen, A.A. (1988) Geophys. Astrophys. Fluid Dyn. **41**, 181
155. van Ballegooijen, A.A., Martens, P.C.H. (1989) ApJ **343**, 971
156. van Driel-Gesztelyi, L., Schmieder, B., Cauzzi, G., et al. (1996) Solar Phys. **163**, 145
157. van Driel-Gesztelyi, L., Wiik, J.E., Schmieder, B., et al. (1997) Solar Phys. **174**, 151
158. van Speybrock, L.P., Krieger, A.S., Vaiana, G.S. (1970) Nat **227**, 818
159. Vekstein, G.E. Priest, E.R. (1992) ApJ **384**, 333
160. Vekstein, G.E., Priest, E.R., Amari, T. (1991) A&A **243**, 492
161. Vršnak, B., Ruždjak, V., Rompolt, B. (1991) Solar Phys. **136**, 151
162. Webb, D.F., Martin, S.F., Moses, D., Harvey, J.W. (1993) Solar Phys. **144**, 15
163. Weiss, L.A., Gosling, J.T., Mc Allister, A.H., et al. (1996) A&A **316**, 384
164. Wilhelm, K., Marsch, E., Dwivedi, B.N., et al. (1998) ApJ **500**, 1023
165. Withbroe, G.L. (1988) ApJ **325**, 442
166. Wolfson, R. (1989) ApJ **344**, 471
167. Wolfson, R., Bhattacharjee, H., Dlamini, B. (1996) ApJ **463**, 359
168. Woo, R.T., Armstrong, J.W., Bird, M.K., Pätzold, M. (1995) ApJ **449**, L91
169. Yoshida, T., Tsuneta, T. (1996) ApJ **459**, 342
170. Zweibel, E.G., Li, H.S. (1987) ApJ **312**, 423
171. Zwingmann, W., Schindler, K., Birn, J. (1985) Solar Phys. **99**, 133

Structuring of the Magnetospheric Plasma by the Solar Terrestrial Interactions

Dominique Fontaine

CETP-CNRS, 10-12 Avenue de l'Europe, F-78140 Vélizy, France,
e-mail: Dominique.Fontaine@cetp.ipsl.fr

Abstract. The existence of a magnetospheric cavity around a planet depends on the interactions of the planet including its atmospheric and magnetic environment with the interplanetary medium. A magnetized planet like the Earth sets a magnetic obstacle against the supersonic super-Alfvénic solar wind flow. The solar wind pressure shapes the magnetosphere, compressing it on the dayside to a few Earth's radii while the nightside tail extends to hundreds of Earth's radii. Away from a homogeneous and constant distribution, very different plasma regions have been identified inside the magnetosphere. Mass and energy transfers with the solar wind are considered as responsible for the magnetospheric plasma structure and dynamics at large-scale as well as for impulsive or transient events. However, these transfer processes remain poorly understood, and reconnection and other working assumptions are presently put forward and developed. Detailed descriptions of the magnetosphere at various complexity levels can be found in textboo ks on space plasma physics. This simplified introduction only aims at proposing keys to get an insight into the structure of the magnetospheric plasma, into a few basic concepts and specific processes at the root of the present understanding and also into questions and issues to be addressed in the future.

1 Introduction

The magnetosphere designates the cavity carved out in the solar wind flow around a planet by the plasmas and fields of planetary origin. By comparison to the field complexity and/or plasma source multiplicity in other planetary magnetospheres, the terrestrial magnetosphere appears extremely simple from this point of view. The only internal plasma source is the Earth's ionosphere. The planetary magnetic field can be described by a dipole with its axis slightly tilted relative to the rotation axis. In the magnetosphere, the dipolar model mainly holds with a very good approximation for the inner regions within typically 10 Earth's radii. Beyond, the solar wind effects contribute to compress it on the dayside and to stretch it along the nightside tail.

The term of cavity, often used for the magnetosphere, refers to the low concentration of the magnetospheric plasma, diluted by one order of magnitude or more relative to the solar wind. Spacecraft observations have in fact identified very different regions, as illustrated in the simplified sketch of

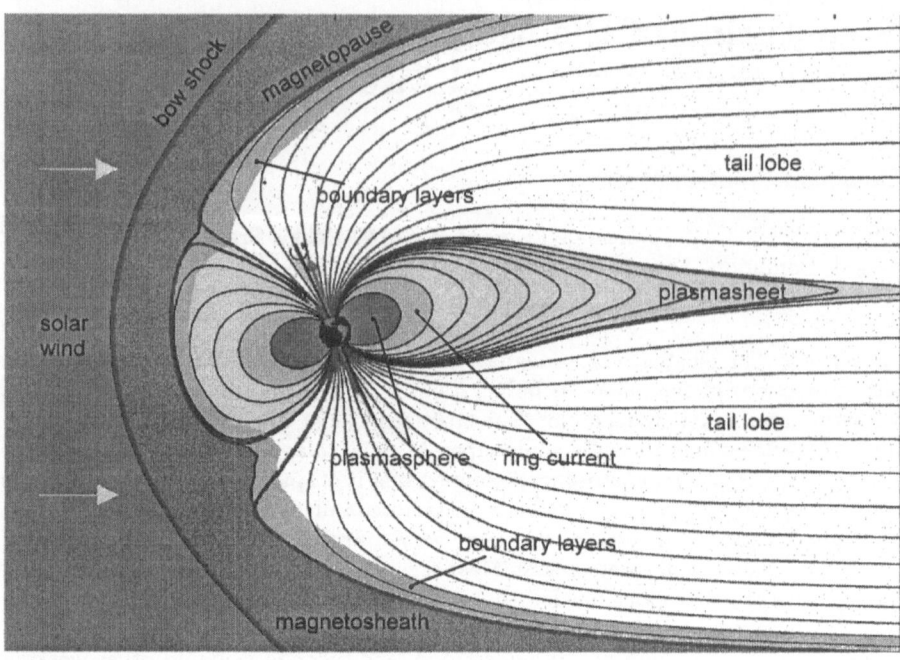

Fig. 1. Cross-section of the magnetosphere in the noon-midnight meridian plane (adapted from personal communication by P. Robert, CETP)

Fig. 1. Just inside the magnetopause, which is the boundary of the magnetosphere with the solar wind, lies a thin boundary layer. It consists of several parts at high or low latitudes but it exhibits everywhere a steep density gradient with the radial distance to the magnetopause: starting from values of the order of $5.10^6\,m^{-3}$ in the solar wind outside the magnetopause, the density decreases by a factor of about 10 across the boundary layer. Its temperature of the order of $10 - 100\,eV$ is comparable to the solar wind values in the magnetosheath, downstream from the bow shock. At larger distances from the magnetopause, the tail lobes are generally considered as empty regions; the plasma becomes extremely rarefied, less than $10^4\,m^{-3}$. Starting from several Earth's radii and extended across the whole magnetotail around the equatorial plane, the plasmasheet concentrates most of the magnetospheric plasma. Relative to the boundary layer, the plasma reaches both weaker densities ($\sim 5.10^5\,m^{-3}$) and much higher temperatures, exceeding the keV, that is to say larger by a factor 100 about. The temperature still increases near the plasmasheet edge and it can reach hundreds of keV or more in the ring current around 6 Earth's radii, which is induced by energetic particle drifts encircling the Earth. Such energetic particles are also detected closer to the Earth in the radiation belts, where they are locally trapped and bounce back and forth between both hemispheres along dipolar field lines. They are embedded in the plasmasphere, the innermost plasma region within a few Earth's radii

about: the dense ($\sim 10^8\,m^{-3}$) and cold ($\sim 1\,eV$) plasma that it contains clearly points to the ionosphere as its direct source.

Without detailing any further, general questions arise out of this simplified description, such as the origin of the different plasma regions observed in the magnetosphere, the nature of the processes governing the density and energy distributions or the relationships between the magnetospheric plasma and the solar wind. Recent and excellent textbooks extensively address a number of questions in space plasma physics, for example *Basic Space Plasma Physics* by Baumjohann and Treumann (1997), *Introduction to Space Physics* edited by Kivelson and Russell (1995) or *Convection and Substorms: Paradigms of Magnetospheric Phenomenology* by Kennel (1995). The present chapter only intends to present a brief and simplified introduction to the magnetospheric plasma, to point out some of the main processes involved in the magnetospheric structure and dynamics, and to give a flavor of recent evolutions in this research field.

The terrestrial magnetosphere results from the interaction of the interplanetary medium with the planet Earth, its atmospheric and magnetic environment. A first approach consists of examining the energetics of the interacting bodies and of sorting out the dominant energy sources (Sect. 2). It demonstrates the role of the magnetic field of planetary origin and of the solar wind which hits on the planetary magnetic obstacle during its expansion in the interplanetary medium at super-sonic super-Alfvénic speeds. Simplified magnetohydrodynamic equilibrium conditions at the magnetopause provide realistic estimates of the magnetosphere's shape and of the solar wind flow around it (Sect. 3). Section 4 briefly analyses the case of the reconnection between the planetary and interplanetary magnetic fields and summarizes some of its fundamental consequences such as the solar wind penetration into the magnetosphere and the generation of a large-scale plasma transport between the different magnetospheric regions. Recent observations have revealed the presence of impulsive and bursty processes at the magnetopause and in the magnetotail, which could be interpreted as reconnection signatures (Sect. 5). These fundamental questions of reconnection, of plasma and energy transfers between the solar wind and the magnetosphere remain poorly understood; they represent the targets of future missions devoted to the study of plasma interfaces like CLUSTER.

2   Energetics of the Interacting Media

The Earth with its magnetic and atmospheric environment is immersed into the interplanetary medium. In order to understand the nature of their interactions, we first start by examining the dominant energy sources stored in both media.

2.1   The Interplanetary Medium at the Earth's Orbit

Among the various components, solar plasma, interstellar neutral gas, cosmic rays and others detected in the interplanetary medium, the main responsible for the interaction with the terrestrial environment is the solar wind. Briefly, the solar wind plasma, with H^+ ions as dominant ion species, is produced and accelerated in the solar corona; beyond several solar radii, it expands at large speeds throughout the solar system. According to the Magnetohydrodynamics theory (MHD), this radial expansion, combined with the solar rotation, results in a spiral configuration of the magnetic field lines. At the Earth's orbit, the observed solar wind parameters typically scale as:

$$\text{Plasma (dominant species } H^+\text{): } n \sim 5.10^6 \, m^{-3}$$
$$T \sim 10 \, eV$$
$$V \sim 400 \, km/s$$
$$\text{Magnetic field amplitude:} \quad B \sim 4 \, nT$$

where n, T, V, B stand for density, temperature, velocity, magnetic field. These typical values provide an estimate of the orders of magnitude of the energy densities stored in the solar wind from the available sources:

$$\text{Ion thermal motion:} \quad P_T \sim 8.10^{-12} \, J.m^{-3}$$
$$\text{Ion bulk flow:} \quad P_D \sim 7.10^{-10} \, J.m^{-3}$$
$$\text{Magnetic field amplitude:} \; P_M \sim 6.10^{-12} \, J.m^{-3}$$

where P_T, P_D, P_M stand for the Thermal, Dynamic and the Magnetic pressures.

The solar wind observed at the Earth's orbit is not stationary but exhibits possibly large variations in density, velocity, pressure. The interplanetary magnetic field can point in any direction and even reverse on short time scales. Intense events like Coronal Mass Ejections (CME), high-speed jets for example, propagate through the solar system and cause strong departures from average values. Therefore, the solar wind energy densities also fluctuate, and the estimates given above only provide orders of magnitude useful to point out the dominant processes. In the solar wind plasma, the dynamic pressure of the bulk flow exceeds the magnetic and thermal pressures by two orders of magnitude. Indeed, these scales express in terms of pressures the more usual relationships known about velocities, stating that the solar wind velocity typically exceeds the sound speed and the Alfvén speed by a factor 10.

2.2   The Earth's Environment:
The Upper Atmosphere and the Ionosphere

As most planets in the solar system, the Earth is surrounded by an atmosphere. Ionized particles are created in the upper layers of the atmosphere by solar radiation. Typically, the solar radiation at UV and X wavelengths below 300 nm does not reach the ground. Its absorption depends on the wavelength, on the atmospheric constituents and on the altitude. The spectrum is relatively complex, but only wavelengths below 150 nm can be responsible for the ionization of neutral atoms and molecules. The ionization processes occur above about 80-km altitude and the produced ionized species form the Earth's ionosphere, embedded in the upper atmospheric layers. At the lower altitudes from 80 km to 160 km, heavy species are created, mainly NO^+ and O_2^+. In the upper ionosphere, O^+ dominates over hundreds of kilometers up to 1000 - 2000 km, and beyond remain only lighter ions like H^+. Of course, the ion concentration fluctuates as a function of the solar fluxes, the solar zenithal angle, or other ionospheric phenomena, but the typical height profiles generally present a peak of the order of 10^{11} to $10^{12} \, m^{-3}$ at about 300 km altitude. By comparison, the neutral concentration exponentially decreases with the altitude from the ground level, but with values of the order of $n_n \sim 10^{15} \, m^{-3}$ at the same altitude, it still exceeds the ionized gas concentration by several orders of magnitude: the ionosphere can be regarded as a weakly ionized gas, embedded in dense atmospheric layers.

Despite the high density of neutral constituents in the upper ionosphere, the collision frequency becomes negligible relative to the gyrofrequency typically above 180 km (called F region), and the charged particle motion is then dominantly controlled by the electromagnetic effects rather than by collisions with neutrals. In presence of electric field, all species, electrons and ions, are drifting across magnetic field lines at the same bulk velocity equal to the electromagnetic velocity ($\boldsymbol{E} \times \boldsymbol{B}/B^2$). This motion, dominant feature in the upper ionosphere, does not produce any current perpendicularly to the magnetic field. Currents are thus constrained to flow in the direction aligned along magnetic field lines.

In the lower ionosphere (region E) typically between 80 km and 180 km, the collisions between ions and neutrals, in particular, become comparable to and compete with the electromagnetic effects. The perpendicular conductivities take significant values and the currents can flow in any direction. In conclusion, field-aligned currents can flow everywhere in the lower as in the upper ionosphere. Their closure by currents perpendicular to the magnetic field can only occur within this lower ionospheric layer, about 100-km thick, that is to say over a very limited altitude range by comparison to the planet's size. Finally, and from an electrical point of view, the ionosphere, ionized gas embedded in the upper atmosphere, can be regarded as a thin spherical conductor encircling the Earth which contributes to close large-scale circuits of currents flowing along magnetic field lines from or toward outer regions of the Earth's environment.

In the ionosphere, different energy sources co-exist due to the presence of the terrestrial magnetic field, the atmospheric neutral gas, the ionospheric plasma and its bulk drift across magnetic field lines. The characteristic values in the ionosphere at 300 km, altitude of the concentration peak, are:

Neutral atmosphere: $n_n \sim 10^{15}\, m^{-3}$
$$T_n \sim 1000\, K$$

Ionosphere (F-region, dominant species: O^+): $n_i \sim 10^{11}\, m^{-3}$
$$T_i \sim 2000\, K$$
$$V_i \sim 1\, km/s$$

Magnetic field amplitude: $B \sim 4.10^4\, nT$

where the subscripts n and i refer to the neutral and ionized components. The energy densities stored in the upper atmosphere and in the ionosphere are estimated for the different sources as follows:

Neutral thermal motion: $P_A \sim 1.10^{-5}\, J.m^{-3}$
Ion thermal motion: $P_T \sim 3.10^{-9}\, J.m^{-3}$
Ion bulk flow: $P_D \sim 1.10^{-9}\, J.m^{-3}$
Magnetic field amplitude: $P_M \sim 6.10^{-4}\, J.m^{-3}$

where P_A represents the Atmospheric thermal pressure. The thermal pressure of the dense atmospheric layers dominate all other terms from the ground level. Then the exponential decrease of the neutral concentration towards higher altitudes causes the decrease of the neutral thermal pressure. At 300-km altitude (F region), it is weaker than the magnetic pressure by almost two orders of magnitude: the plasma behavior has split off from the neutral dynamics to become governed by the magnetic field. It is equivalently expressed in other words that, in the ionospheric F-region, the charged particle gyrofrequency exceeds the collision frequencies between species, and in particular with neutrals. The energy density stored in the cold, slow and weakly ionized ionospheric plasma appears quite negligible relative to the energy densities from the atmospheric and magnetic sources.

2.3   The Earth's Environment:
The Terrestrial Magnetic Field and the Magnetosphere

Above the ionosphere, in the magnetosphere, all energy densities are expected to decrease with the distance from the Earth. The neutral concentration, exponentially decreasing from the ground level, becomes extremely weak; the plasma, mainly H^+, is very diluted; and the magnetic field also decreases according to the dipolar power law as r^{-3}. The different plasma regions existing in the magnetosphere are described in Sect. 1. Their characteristics vary, but to give orders of magnitude, we focus here on the plasmasheet, the extended region around the equatorial plane which plays a dominant role in the magnetosphere. Relative to the ionosphere or the solar wind, the plasmasheet contains much more diluted and energetic plasma with the following typical values at ten Earth's radii:

Plasma (dominant ion species H^+): $n \sim 5.10^5 \, m^{-3}$
$$T \sim 1 \, keV$$
$$V \sim 25 \, km/s$$
Magnetic field amplitude: $\qquad\qquad$ $B \sim 40 \, nT$

The energy densities are estimated from the different energy sources as:

Ion thermal motion: $\qquad$ $P_T \sim 2.10^{-11} \, J.m^{-3}$
Ion bulk flow: $\qquad\qquad$ $P_D \sim 5.10^{-13} \, J.m^{-3}$
Magnetic field amplitude: $P_M \sim 6.10^{-10} \, J.m^{-3}$

As expected, all magnetospheric pressures are smaller than in the iono-sphere, but the magnetic field still remains the dominant energy source. In particular, it largely exceeds the pressure of the plasma itself, more energetic but also much more diluted than in the ionosphere. Diluted and governed by the magnetic field, the magnetospheric plasma exhibits the main feature of being collisionless. Finally, these estimates lead to the conclusion that, above the dense atmospheric layers dominated at low altitude by the neutral ther-mal pressure, the magnetic field controls the plasma dynamics in the upper ionosphere and beyond, in the whole terrestrial environment. It is playing a central role in the interaction of the Earth with the interplanetary medium.

For comparison, the energy density estimates in the terrestrial environ-ment and in the solar wind are collected in Table 1.

Table 1. Typical orders of magnitude of energy densities stored in the ionosphere (at a reference altitude of 300 km), in the magnetosphere (at about 10 Earth's radii) and the solar wind (at Earth's orbit).

Energy sources:	Energy densities (expressed in $J.m^{-3}$) in:		
	Ionosphere	Magnetosphere	Solar Wind
Neutral thermal motion	$\sim 1.10^{-5}$		
Ion thermal motion	$\sim 3.10^{-9}$	$\sim 2.10^{-11}$	$\sim 8.10^{-12}$
Ion bulk flow	$\sim 1.10^{-9}$	$\sim 5.10^{-13}$	$\sim 7.10^{-10}$
Magnetic field	$\sim 6.10^{-4}$	$\sim 6.10^{-10}$	$\sim 6.10^{-12}$

The orders of magnitude in Table 1 suggest a first picture of the solar ter-restrial interactions. In the interplanetary medium, the solar wind is mainly responsible for interacting with the Earth's environment. Its main energy source is stored in its bulk flow drifting at supersonic and super-Alfvénic speeds. As most planets, the Earth can oppose its gaseous envelop. Starting from the ground level, the atmospheric thermal pressure dominates in the dense layers of the neutral atmosphere. It rapidly decreases with altitude, and at the altitude of the ionization peak in the ionosphere (about 300 km), the magnetic pressure has not only exceeded the pressure of the ionized gas but also the pressure of the neutral gas, which is the domimant constituent at those heights. At larger distances, all pressures decrease in the very tenuous magnetospheric plasma. The dominant energy source is again stored in the magnetic field, even if the magnetic pressure is divided by several orders of magnitude relative to its value in the ionosphere. In other words, the Earth sets a magnetic obstacle against the dynamic pressure exerted by the supersonic super-Alfvénic solar wind. In Table 1, the comparable orders of magnitude between the magnetospheric magnetic pressure and the solar wind dynamic pressure intuitively suggests that a balance can be achieved.

3 Fluid Approach of the Solar Wind/Magnetosphere Interaction: The Concept of Closed Magnetosphere

From the dominant energy sources estimated in Sect. 2, the interaction between the interplanetary medium and the planet Earth can simply be approached by the model of a monokinetic jet of solar plasma flowing at supersonic, super-Alfvénic speeds against a magnetic dipole. The opposite case of the supersonic aircraft flying through the atmospheric gas at rest is well known to produce a bow shock upstream from the obstacle. Considered in the reference frame of the obstacle, the interaction of the supersonic solar wind with the planetary dipole similarly produces a bow shock upstream from the magnetosphere. The bow shock contributes to decelerate at subsonic and sub-Alfvénic speeds the solar wind, which can then turn around the obstacle. In terms of energy budget, part of the energy initially stored in the solar wind bulk flow upstream from the bow shock is converted downstream into thermal and magnetic energy. Afterwards, the solar wind progressively accelerates again along the obstacle flanks, recovering supersonic super-Alfvénic speeds.

The interaction between the solar wind and the Earth is also responsible for the existence of the magnetopause, boundary that separates the interplanetary medium and the solar wind flow from the terrestrial environment controlled by the planetary magnetic field. The solar wind flowing outside the magnetopause along its flanks does not penetrate into it; conversely, the planetary magnetic field remains confined inside the magnetosphere and does not diffuse into the interplanetary medium. In a first approximation, the magnetopause is an equilibrium boundary which achieves the balance between the dynamic pressure of the solar wind and the magnetic pressure prevailing in the magnetospheric vacuum.

A brief description of individual particle motions near the magnetopause enlightens the microphysic processes invoed in the solar wind - magnetosphere interaction. Ideally, the solar wind particles experience a *specular reflection* at the magnetopause, as illustrated in Fig. 2. Let us assume that the oncoming solar wind particles can cross the magnetopause, thus they come under the influence of magnetospheric magnetic field. They immediately describe a cyclotron rotation around magnetic field lines, which brings them back at the magnetopause before having completed a full turn. They escape again into the interplanetary medium at velocities symmetric of incident velocities relative to the normal at the magnetopause, as for a mirror reflection. Finally, the solar particles cannot penetrate the magnetosphere over distances much larger than twice their Larmor radius before turning back, and the net mass flow across the boundary is zero. Indeed, the particle Larmor radii give a scale for the magnetopause thickness: physically, it represents the distance necessary for the interface layer to achieve an equilibrium with jump conditions in plasmas and fields on both sides. Inside the magnetopause layer, the electrons and ions describe their cyclotron motion in an opposite direction, which generates a current. Regarded as a current layer, the magnetopause is playing the role of a Faraday cage for the magnetosphere. This explains in particular its shielding role between the magnetospheric and interplanetary media. Finally, the magnetopause currents produce a magnetic field which adds to the initial magnetic field of planetary origin.

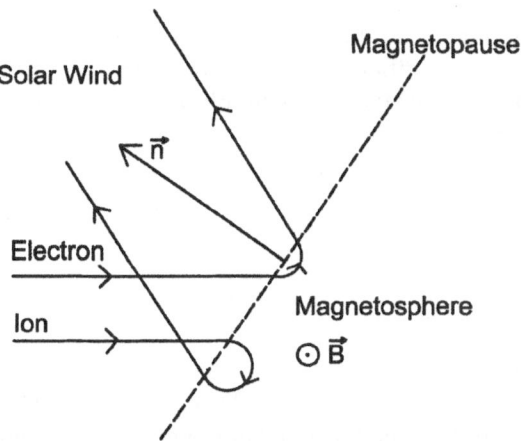

Fig. 2. Sketch illustrating the specular reflection of solar wind particles at the magnetopause

The equilibrium conditions at the magnetopause can be computed from MHD and from the general conservation laws in fluids. In presence of discontinuities, these conservation laws lead to jump conditions between both sides of the discontinuity. In first approximation, the magnetopause can be regarded as a tangential discontinuity in equilibrium between the solar wind dynamic pressure and the magnetospheric magnetic pressure. No mass flux can flow across it, and with the neglect of the interplanetary magnetic field, the normal component of the magnetic field vanishes across the boundary. The pressure balance on both sides can be simply calculated from the Rankine-Hugoniot jump equations (see Heyvaerts, this issue):

$$n\, 2\, \rho_m\, V_{SW}^2\, \cos^2 \chi = n\, \frac{B^2}{2\mu_0} \qquad (1)$$

where ρ_m is the solar wind mass density. The angle χ of the oncoming particle velocity V_{SW} with the normal n to the magnetopause depends on the magnetopause geometry. The left-hand member represents the total impulsion flux on the solar wind side due to the bulk motions of both oncoming and reflected particles. On the magnetospheric side (right-hand member), the only contribution to the impulsion flux comes from the terrestrial magnetic field B. Finally, the equilibrium conditions at the magnetopause relate the amplitude of the tangential magnetic field (since the normal component is equal to zero, $B_n = 0$) to the solar wind parameters:

$$B_{mp}^2 \equiv B_t^2 = 4\,\mu_0\,\rho_m\,V_{SW}^2\,\cos^2\chi \qquad (2)$$

Equation 2 does not depend on the location of the magnetopause relative to the terrestrial dipole but includes an implicit relation to the magnetopause shape through the angle χ.

Finally, the search for the magnetospheric magnetic field distribution in the magnetosphere leads to solve a differential equation with conditions at free boundaries. With the approximation of negligible current distribution inside the magnetospheric cavity, the problem reduces to search for a scalar potential γ such that:

$$B = -\nabla \gamma \qquad (3)$$

with the boundary conditions:

- at the Earth's surface. This is a fixed boundary, and the terrestrial magnetic field distribution on the ground is well known,
- at the magnetopause. This is the free boundary of the problem, because of its unknown location and shape. The normal component of the magnetic field cancels out ($B_n = 0$) and the tangential component obeys (2) at the magnetopause.

A simplified approach. This problem is often called after Chapman and Ferraro who were the first to propose a solution consistent with the present state of the art (Chapman and Ferraro, 1931). Indeed, they addressed a simplified question, illustrated in Fig. 3: briefly, they assumed that the terrestrial dipole axis was perpendicular to the Sun-Earth axis and that the magnetopause could be assimilated to an infinite plane also perpendicular to the Sun-Earth axis. It thus defines two half spaces with the solar wind on one side and the magnetosphere on the other one. Of course, the magnetic field

distribution in the magnetospheric half-space results from the interaction of
the solar wind with the terrestrial dipole, but it can be indirectly computed.
The magnetic image method predicts that this magnetospheric distribution
would be identical to the distribution produced by the interaction of two symmetric dipoles: the real terrestrial dipole (M) plus a virtual dipole (M') of
equal magnetic moment and located at the symmetric distance relative to the
planar magnetopause along the Sun-Earth axis. Such a configuration satisfies the boundary condition that the normal component of the magnetic field
cancels out at the magnetopause ($B_n = 0$). Although extremely simplified,
this picture provides interesting estimates.

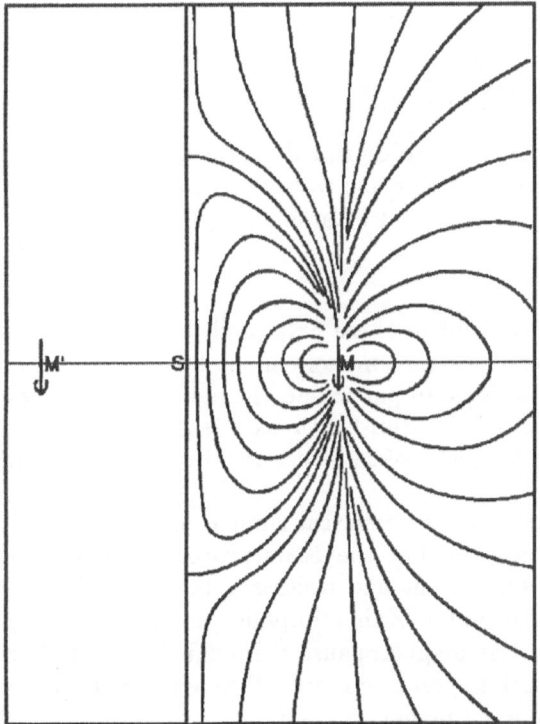

Fig. 3. Sketch representing the terrestrial dipole (M) and the image dipole (M') in
the noon-midnight meridian plane (adapted from the original sketch by Chapman
and Ferraro (1931))

For example, let us consider the sub-solar point S at the magnetopause
where the planar magnetopause intersects the axis Sun-Earth (Fig. 3). It is
located just in the middle between the two dipoles and the total magnetic

field is simply twice the field produced by the terrestrial dipole at the distance L(S):

$$B(S) = 2\,\frac{B_0}{L(S)^3} \tag{4}$$

where B_0 is the amplitude of the magnetic field at the Earth's surface at the equator ($B_0 \simeq 4000\,nT$) and L(S), the geocentric distance of the point S, expressed in Earth's radii, is also the geocentric of the magnetopause L_{mp} along the Sun-Earth axis. These simple considerations quantify the magnetic field increase predicted and observed at the nose of the magnetopause.

Moreover, the combination with the equilibrium condition (2) at the magnetopause provides an estimate of the magnetopause geocentric distance L_{mp}, which remained unknown:

$$L_{mp} \equiv L(S) = \left(\frac{B_0^2}{\mu_0\,\rho_m\,V_{SW}^2\,\cos^2\chi}\right)^{1/6} \tag{5}$$

For a solar wind flow approximately aligned along the line Sun-Earth ($\chi \simeq 0$), and with the orders of magnitude given in Sect. 1, this relation predicts the magnetopause location at $L_{mp} \simeq 10$ Earth's radii, in good agreement with the observations.

The present models. Successful in these first predictions of the magnetopause location, the Chapman-Ferraro model cannot be exploited any further because of obvious limitations. In particular, the assumption of a planar magnetopause perpendicular to the Sun-Earth axis is not verified, except maybe near the nose of the magnetopause, it cannot describe the magnetospheric tail.

More recent models make use of numerical simulations to compute the magnetopause shape and location, but the basic procedure remains practically unchanged and the subsequent and necessary improvements do not actually modify the physical meaning of this simplified approach. Briefly, the equilibrium condition (2) at the magnetopause is modified to include a parameter κ which describes various conditions and efficiency variations for the specular reflection along a curved magnetopause:

$$B_{mp}^2 \equiv B_t^2 = \kappa\,\mu_0\,\rho_m\,V_{SW}^2\,\cos^2\chi \tag{6}$$

Then, once the three-dimensional shape of the magnetopause is computed, the fluid theories predict the solar wind flow distribution towards the obstacle and its deviation around it. Results from one of the first quantitative three-dimensional studies, then exploited for years, is illustrated in Fig. 4 (Spreiter et al., 1966). The steamlines, represented by the dashed lines, show the flow deviation experienced by the solar wind downstream from the bow shock (outer boundary) around the magnetopause (inner boundary). The fluid parameters estimated along these streamlines allow to quantify the flow deceleration, compression and heating just downstream from the shock at the nose of the magnetopause and then its progressive acceleration and cooling along the magnetopause flanks.

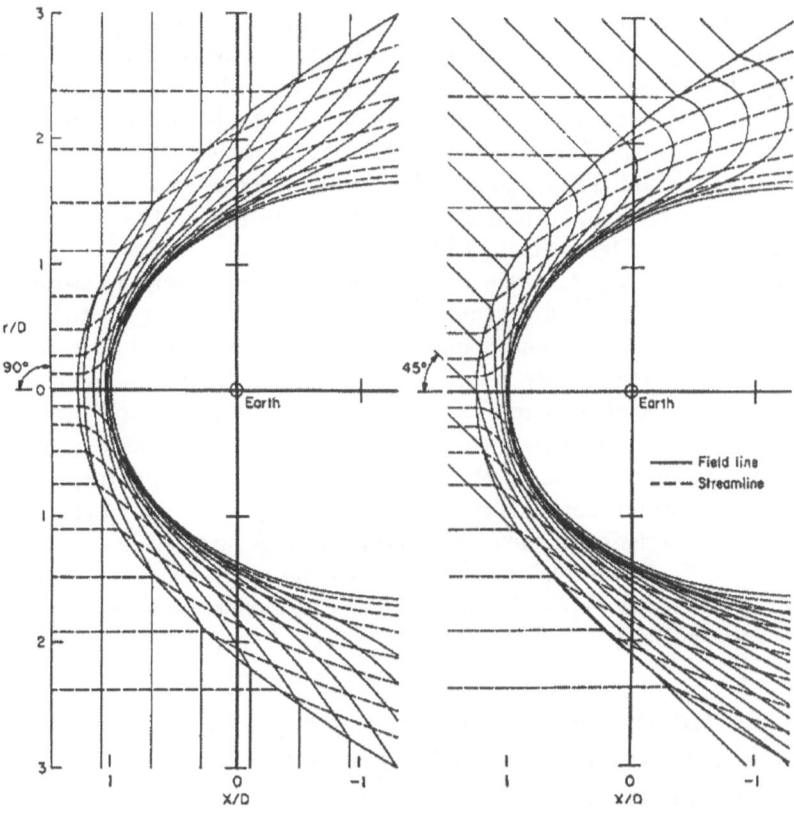

Fig. 4. Solar wind flow around the magnetopause (Spreiter et al., 1966)

This approach does not take into account the interplanetary magnetic field because its pressure is negligible relative to the solar wind dynamic pressure. Indeed, the magnetic field is simply transported with the solar wind flow. Its distribution can be computed from the Maxwell's equations with the approximation of frozen field (see Heyvaerts, this issue):

$$\frac{\partial B}{\partial t} = \nabla \times (V_{SW} \times B) \tag{7}$$

The magnetic field lines (solid lines) are represented in Fig. 4 for two initial conditions with an upstream magnetic field perpendicular to the flow (left) or with a 45° tilt (right). The magnetic field lines bend after the bow shock, drape around the magnetopause and pile up at the nose of the magnetopause, thus increasing the magnetic flux in this region. In other words, the solar wind dynamic pressure upstream from the bow shock is converted downstream into thermal and magnetic pressure. The agreement between the simulated and measured parameters brings the evidence of the physical reality contained in this simple description of the solar wind - magnetosphere interaction. Numerical efforts have then substantially developed. Nowadays, the computer power allows solving self-consistently the equations of the three-dimensional magnetohydrodynamics (MHD) and Maxwell's equations from conditions in the upstream solar wind and at the planetary surface. Both free boundaries, the bow shock and the magnetopause, appear naturally in the simulation box. These global MHD modelings can also take into account more complex plasma, current and field configurations in the magnetosphere (see for example, Ogino et al., 1986).

In conclusion, the analysis of the dominant energy sources stored in the interacting bodies, Earth and interplanetary medium, has led to a successful description of the magnetopause: it appears as an equilibrium boundary, shielding the Earth and its magnetic environment, the magnetosphere, from the direct impact of the super-sonic and super-Alfvénic solar wind, which is deviated around it. However, this conclusion admits of some departures and does not explain some observations. For example, the magnetosphere is not an empty cavity but contains different plasma regions. Some of them present similar characteristics to the solar wind plasma. A large-scale electric field is observed in the dawn-dusk direction across the magnetosphere. And finally, the auroral activity above the high-latitude regions is found to be correlated to the solar activity.

4   The Interactions Between Interplanetary and Planetary Magnetic Fields: The Concept of Convection

The observed presence of fields, plasmas, and auroral activity in the terrestrial magnetosphere requires to improve our simplified description of a fast solar wind flow hitting the planetary dipole. Hereafter, we will focus on the magnetized feature of the solar wind plasma. With negligible effects in terms of energetics (see Sect. 2), the interplanetary magnetic field may however play a particular role in the interaction between the solar wind and the magnetosphere. This can be roughly illustrated in the following way. At the Earth's orbit, the direction and the amplitude of the interplanetary magnetic field fluctuates due to the solar rotation, and the polarity reverses depending on the Earth's location relative to the solar neutral sheet. It may happen that, at some times and at some places at the magnetopause, the interplanetary magnetic field becomes equal and opposite to the terrestrial magnetic field. Their combination would produce a reconnection figure with an X-point where the total field cancels out: the solar wind plasma could then freely penetrate into the magnetosphere.

Again, the interaction between the planetary and interplanetary magnetic fields can be quantitatively approached with a simplified geometry, ignoring in a first step the dominant dynamic pressure in the solar wind. Let us consider the simple interaction of the terrestrial dipole with an interplanetary field, constant, homogeneous and parallel to the dipole axis. Although simplified, this case basically contains the physical meaning of more realistic and complex field configurations.

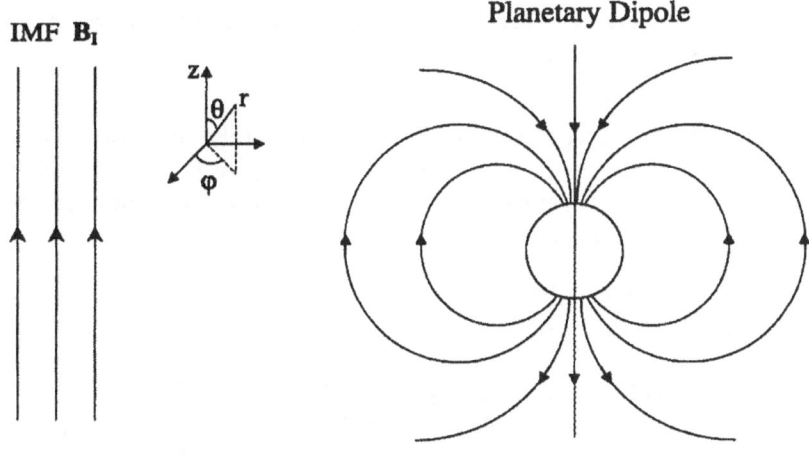

Fig. 5. Simplified representation of a northward Interplanetary Magnetic Field B_I and the terrestrial dipole

Let us define the z-axis as the dipole axis, as illustrated in Fig. 5. In the spherical coordinate system with the origin at the Earth's center, r is the radial distance, θ the colatitude counted from the z-axis (north), and φ the azimuthal angle in the equatorial plane perpendicular to the z-axis. The components of the dipolar magnetic field B express as:

$$B_r = -2 \cos \theta \, B_0 \left(\frac{R_E}{r} \right)^3$$

$$B_\theta = - \sin \theta \, B_0 \left(\frac{R_E}{r} \right)^3$$

$$B_\varphi = 0$$

where R_E represents the Earth'radius and B_0 the magnetic field amplitude at the Earth 's surface at the equator. The interplanetary magnetic field $\boldsymbol{B_I}$ is simply considered as constant along the z-axis: $\boldsymbol{B_I} = B_I \, \boldsymbol{e_z}$

The magnetic field lines resulting from the superposition of both magnetic fields are axisymmetric since both fields do not depend on the φ component. Their equation in any meridian plane (at a constant φ) can be written as:

$$+ \frac{B_I}{2} \left(\frac{r}{R_E}\right)^2 \sin^2\theta - B_0 \left(\frac{R_E}{r}\right) \sin^2\theta = constant \tag{8}$$

At short distances from the Earth, the dipolar contribution dominates (second term on the left-hand side), the magnetic field lines are approximately dipolar, and conversely at large distances, the interplanetary field dominates (first term).

Case of a northward IMF. In the case of a purely northward interplanetary magnetic field, the constant B_I is positive. Since B_0 is also positive, the left-hand member can vanish, and the choice of a null constant at the right hand member defines a particular field line, independent of θ:

$$\forall\varphi, \, r = R_E \left(\frac{2\,B_0}{B_I}\right)^{1/3} \tag{9}$$

The set of field lines, obeying (9) for any angle φ, defines a sphere around the Earth, which separates closed field lines of dipolar origin inside from open field lines in the interplanetary medium outside. No connection exists between both media, separated by this spherical magnetopause. This is the so-called model of the "dipole-in-sphere", illustrated in the left part of Fig. 6, which has led to the concept of "closed magnetosphere". The solar dynamic pressure additionally contributes to distort the initial sphere into an egg-shaped magnetosphere, compressed in the Sun direction and extended in the opposite direction, but it will not change the essential point that the magnetosphere is closed. This situation is equivalent to the model described in Sect. 3 with a magnetosphere fully insulated against any exchange with the interplanetary medium. This is the only situation where a closed configuration can be predicted. All other orientations of the interplanetary magnetic field would behave more or less similarly to the case of a southward interplanetary magnetic field, although the geometry could be much more complex.

Case of a southward IMF. A southward interplanetary magnetic field expresses as: $\boldsymbol{B_I} = -B_I \, \boldsymbol{e_z}$, and the left-hand side in the magnetic field line equation (8) becomes always negative. With this sign change, the circular field lines no longer exist. The new topology is illustruted in the right part of Fig. 6. In addition of the open field lines (A) in the interplanetary medium

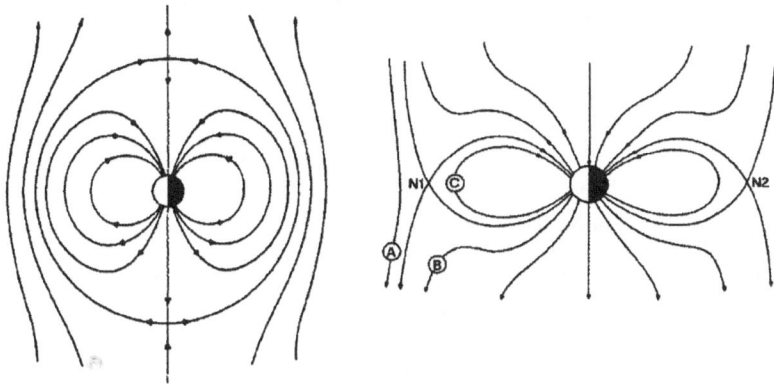

Fig. 6. Superpersition of a dipolar magnetic field with a constant northward (left) or southward (right) field, adapted from Stern (1977)

and of the closed dipolar field lines (C) in the inner regions, appears above the polar caps a new class of field lines (B) with one footprint connected to the planet and the other end immersed in the interplanetary medium. The total magnetic field vanishes at all points such that:

$$\forall \varphi, \ \theta = \frac{\pi}{2}, \ and \ r = R_E \left(\frac{B_0}{B_I} \right)^{1/3} \tag{10}$$

They form, in the equatorial plane of the magnetosphere ($\theta = \pi/2$), a circle of points where the planetary and interplanetary magnetic fields are exactly equal and opposite, resulting in a total field equal to zero. Two of these points N1 and N2 are represented in the meridian cross-section of Fig. 6 perpendicular to the equatorial plane. At all points of this so-called "neutral line", an initially open interplanetary magnetic field line reconnects with an initially closed planetary field line, as illustrated for N1 and N2. Again, the dynamic pressure exerted by the solar wind is expected to distort the circular neutral line into an oval shape, coming much closer to the Earth on the day-side (about 10 R_E) than downtail (possibly one or several hundreds of Earth's radii). This asymmetric configuration is displayed in Fig. 7 which represents a cross-section of the magnetosphere in the noon-midnight meridian plane. Fig. 7 also illustrates the consequences of the reconnection processes on the particle and field dynamics.

The merging of a southward interplanetary magnetic field line with a closed terrestrial field line (both labeled 1) forms two new field lines (labeled 2) with one footprint connected to the polar cap of one hemisphere and the other end immersed in the interplanetary medium. Their initially stiff curvature is getting smoother as they are transported downtail over the polar caps with the solar wind (labels 3 to 6). When they reach the opposite neutral point at the nightside magnetopause, they reconnect (in 7) and produce two field lines (both labeled 8): an open magnetic field line, then dragged

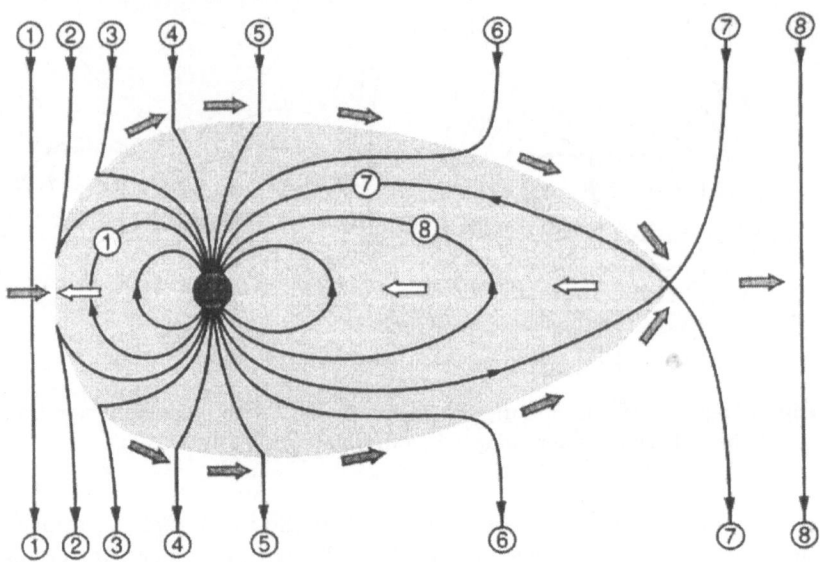

Fig. 7. Sketch of the reconnection model initially proposed by Dungey (1961). (The figure has been adapted from Baumjohann and Treumann (1997))

downstream by the solar wind flow, and a closed field line with both ends connected to the Earth and stretched on the nightside. This stretched field line is then transported Earthward towards the dayside magnetopause, where it contributes to balance the magnetic flux previously depleted by the initial dayside merging process. The same cycle can resume.

With the magnetic field frozen into the plasma, the solar wind plasma is also involved in the reconnection processes between planetary and interplanetary fields. Solar wind particles initially gyrating around an interplanetary field line can penetrate the magnetosphere once the field line has merged with a closed magnetospheric field line (and vice versa for magnetospheric particles). The reconnection processes are responsible for plasma exchanges between solar wind and magnetosphere: this is the concept of open magnetosphere, first proposed by Dungey (1961).

Reconnection has also important implications regarding the plasma transport in the magnetosphere. The plasma contained in newly reconnected magnetic flux tubes at the dayside magnetopause is transported downtail over both polar caps. When the tubes reach the distant nightside reconnection point, part of the plasma escapes into the solar wind with newly formed interplanetary flux tubes. The other part, trapped on closed field lines, is transported sunward toward the planet and the dayside magnetopause (fig. 7). Taking into account the 3-D structure of the magnetosphere, this plasma circulation in the magnetosphere describes two vortices, from the dayside magnetopause above each polar cap toward the nightside tail and then back to the dayside on closed field lines: it is called "convection", because these vortices look like the flow cells generated in a fluid by a heat source (although no heating process is involved here). In the reference frame of the Earth, this magnetospheric plasma flow is equivalent to an electric field, the so-called convection electric field, directed from dawn to dusk across the magnetosphere. Measured by spacecraft, or from ground-based instruments at the field line footprints, the convection electric field is typically of the order of 0.2 mV/m in the magnetosphere, corresponding to a total dawn-to-dusk potential drop across the magnetosphere of 50 kV. By comparison, the antisunward flow of the solar wind at 400 km/s in the interplanetary magnetic field (4 nT at the Earth's orbit) is equivalent to an electric field of 1.6 mV/m in the terrestrial reference frame, resulting in a potential drop of the order of 400 kV over a distance equivalent to the tail diameter ($\sim 40\,R_E$). These figures indicates that roughly 10 % of the magnetic flux carried by the solar wind is transferred inside the magnetosphere.

In conclusion, the reconnection processes between the terrestrial and interplanetary magnetic fields are responsible for particle exchanges between the magnetosphere and the solar wind, and in particular for plasma entries from the solar wind into the magnetospheric cavity. In addition, and this is another important consequence, they also drive a large-scale plasma flow inside the magnetosphere: the convection motion describes two vortices with a tailward flow from the dayside magnetopause over the poles and a sunward return in the inner magnetosphere. In this model of the reconnecting magnetosphere, the interaction between the solar wind and the magnetosphere is assimilated to a dynamo which generates the convection electric field and drives the plasma inside the magnetospheric cavity (Cowley, 1982). From a competing or complementary point of view, Axford and Hines (1961) suggested the fluid viscosity at the magnetopause at the origin of the convection motion. Briefly, this model was initiated from observations near the magnetopause of the plasma bulk velocities, which, from solar wind values at the magnetopause, progressively decrease across the boundary layers. However, the estimates of the resulting magnetospheric transport show that, if the fluid viscosity at the magnetopause may contribute to the convection, it cannot account by itself for the observed fields.

Large-scale electrodynamics in the magnetosphere and auroras in the ionosphere. The penetration of solar wind plasma and its transport in the whole magnetospheric cavity contribute to create plasma regions well identified in the magnetosphere. Just inside the magnetopause, the boundary layers are essentially populated by particles of solar origin that have entered the magnetosphere. They exhibit characteristics similar to the solar wind in the magnetosheath downstream the bow shock, with densities of the order of several $10^6\,m^{-3}$, and temperatures of several tens to hundreds of electron-Volts. The layers on the dayside and on the magnetosphere's flanks are generally called the Low-Latitude Boundary Layer (LLBL), while on the nightside lies the High-Latitude Boundary Layer (HLBL), commonly called the mantle.

The convection transport drives these boundary layer particles to large distances in the nightside tail, typically to hundreds of Earth's radii. Part of them is lost and returns to the solar wind, while others accumulate around the equatorial plane of the magnetotail and form the plasmasheet, particle reservoir permeated by closed magnetic field lines with both ends connected to the planet and a very stretched configuration in the tail. As already mentioned these stretched flux tubes filled with plasmasheet plasma are transported sunward by convection. During their motion, their volume severely decreases, their inner pressure increases, and the plasma is heated: for example, at ten Earth's radii, the plasmasheet temperature has increased to a few keV, for a density of about $10^5 \, m^{-3}$, weaker than in the boundary layers. Closer to the Earth, the planetary rotation deviates the sunward convection drift and first drives the plasma around the Earth before it reaches the dayside magnetosphere. Consequently, the plasmasheet particles cannot approach the Earth closer than several Earth's radii. Near the inner edge, the plasma has still gained energy, and particles with hundreds of keV have been detected in the ring current and in the radiation belt (see Sect. 1). Indeed, in inhomogeneous magnetic configurations involving curvatures and gradients like the terrestrial dipole, the particles additionally experience magnetic drifts around the Earth perpendicularly to the magnetic field. Negligible for low-energy particles, only driven by convection, the gradient and curvature drifts dominate the motion of energetic particles, which become trapped on closed orbits around the Earth. As ions and electrons are drifting in opposite directions, they generate a net transverse current, the ring current. Finally, the innermost magnetospheric region, the plasmasphere, does not remain empty, but its population is not related to the solar wind. It is filled with cold particles of ionospheric origin that have escaped along magnetic field lines and are drifting on closed orbits around the Earth. This cold plasmaspheric plasma, close to the Earth, is predominantly dragged by the co-rotation with the Earth, which largely exceeds any convection or magnetic drift effect.

After having completed this brief description of the magnetospheric populations, we come back here to the nightside plasmasheet. During their sunward transport, the plasmasheet flux tubes injected from the distant tail are losing particles by collisions with the upper atmospheric layers: this is the process of "precipitation" into the ionosphere. Excited by collisions, the atmospheric constituents then relax by emitting these beautiful radiations, dominated by the oxygen green line and visible in the high-latitude sky: the *aurora borealis and australis*. As explained above, the plasmasheet flux tubes, first drifting earthward from the distant tail, are then deviated around the Earth toward the dayside magnetopause. The continuous precipitation at their footprints traces luminous paths in the atmosphere which form these two auroral belts permanently present around the poles. The solar origin of the particles responsible for auroras explains the observed correlations of the location, width or intensity of the auroral ovals with the solar activity.

Spectacular luminous displays, the auroral ovals also mark the location of the most intense currents and fields detected in the terrestrial ionosphere. This enhanced electrodynamics results from close couplings between the auroral ionosphere and the plasmasheet. They are achieved via the magnetic field lines embedded into the dilute energetic plasmasheet plasma and connected at their ends to the dense and cold ionosphere. The magnetic field lines enable plasma exchanges between the ionosphere and the magnetosphere, such as the precipitation of plasmasheet particles into the ionosphere, or conversely, the escape of ionospheric particles into the magnetosphere. They drive currents, and these field-aligned currents ensure the important role to connect and to close currents flowing inside the thin ionospheric conductive shell around the Earth with the plasmasheet curents across the magnetospheric tail: a large-scale current circuit is built between the ionosphere and the plasmasheet. Electric fields are also transmitted along the magnetic field lines between both regions. All these couplings between the auroral ionosphere and the plasmasheet and the feedback effects between both regions contribute to enhance the auroral electrodynamics.

In summary, the dynamic pressure, dominant energy source stored in the solar wind, does govern the solar wind motion and its interactions with the terrestrial dipole. Fluid theories successfully predict the formation and the shape of the bow shock and of the magnetopause, the solar wind flow and magnetic configuration around the obstacle, etc... The contribution of the solar wind magnetic pressure only plays a secondary role in terms of energetics. However, the reconnection processes that take place between the terrestrial and interplanetary magnetic fields have huge consequences for the initially empty magnetospheric cavity. First designed by Dungey (1961), this concept of a reconnecting magnetosphere successfully explains various features observed in the terrestrial magnetosphere, mainly:

- the penetration of solar wind plasma inside the magnetosphere,
- the generation inside the magnetospheric cavity of a large-scale plasma circulation or "convection", which contributes to transport the plasma and to populate different regions of the magnetosphere,
- the ultimate formation of two auroral ovals around the poles, and their dynamics correlated to the solar activity.

The global dynamics of the solar wind/magnetosphere interactions is approached quantitatively from fluid models. Of course, the physical description and the simulation of the reconnection processes themselves remain out of their scope because the fluid approximations are not satisfied near the reconnection sites. However, they provide a useful tool to describe or to predict the large-scale distribution of the magnetospheric plasmas and fields, the plasma transport by convection, the dynamical responses to solar wind variations, the development and propagation of localized or time-dependent events. (Harel et al., 1981; Peymirat and Fontaine, 1994).

5   An Impulsive Reconnection?

From this global understanding of the magnetospheric structure and dynamics, further complexities can be pointed out. Strongly structured by the planetary magnetic field and by internal plasma regions closely connected to the ionosphere on one hand, the magnetosphere experiences on the other hand the effects of the solar wind. The high solar wind variability, as observed at the Earth's orbit, questions the stability of the magnetopause, equilibrium interface between the solar wind and the magnetosphere, and consequently the stability of the plasma regions inside the magnetospheric cavity. The interplanetary magnetic field is playing a central role in reconnection processes and its variability is expected not only to modify their nature or their dynamics but also to alter the subsequent penetration of solar particles and the convection motion inside the magnetosphere.

Observations from spaceborne or ground-based instruments have revealed the occurrence near the magnetopause and in other magnetospheric and ionospheric regions of small time-scale, transient or bursty events, which can be interpreted as the signature of impulsive reconnection. This is briefly illustrated in the two following examples.

Flux Transfer Events (FTE). Among the best-known small-scale transient events, the so-called Flux Transfer Events (FTE) were first discovered at the dayside magnetopause by Russell and Elphic (1979). In addition to their characteristic bipolar magnetic signature, both energetic particles of magnetospheric origin and lower-energy particles from the solar wind are simultaneously detected in FTEs. A scenario often suggested at the origin of FTEs assumes the transient formation of an X-type reconnection point at the magnetopause as illustated in Fig. 8: a terrestrial magnetic flux tube, initially closed, opens and merges with an interplanetary magnetic flux tube into one single flux tube filled with a mixed population as observed. Generally, the FTEs do not occur as isolated events near the dayside magnetopause, but they rather follow each other quasiperiodically. Their statistical recurrence rate varies between about 2 min. to 16 min., with a mean value of 8 min. (Lockwood et al., 1993). Several FTE models, involving one isolated reconnected flux tube or more complex reconnection pictures with multiple tubes more or less twisted, are presently competing. However, most of them agree on possible reconnection sites located on the dayside magnetopause near the subsolar region and preferably in presence of a southward IMF component. These are only favorable conditions and they admit large departures. From the reconnection site, the newly reconnected tubes, with one end immersed in the solar wind and the other one connected to the planet, are then transported antisunward above the polar caps.

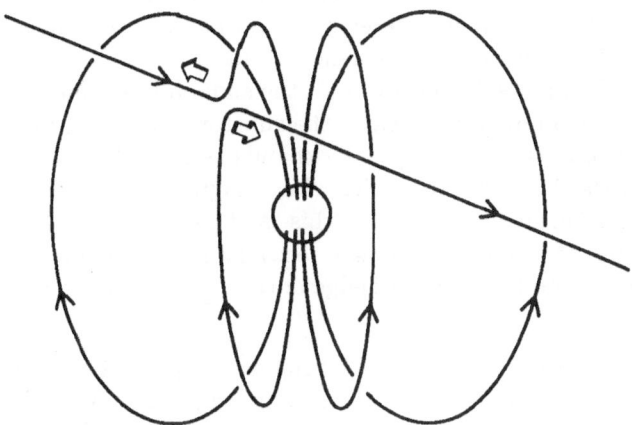

Fig. 8. Sketch illustrating the reconnection between an oblique interplanetary magnetic field line and a dipolar magnetic field line (adapted from Onsager and Fuselier, 1994)

FTEs are predicted to produce distinct signatures at their ionospheric footprints (in the region of the polar cusps) after a time delay corresponding to the Alfvén travel time from the reconnection site to the conjugated ionosphere ($\sim$ 2 min.) (Southwood, 1987). Indeed, in the high-latitude ionosphere near noon, ground-based optical instruments have observed series of polward moving auroral forms, created by precipitating particles in an energy range consistent with the solar wind plasma, while radars have detected sporadic and intense bursts of plasma flows. Their repetition period comparable to FTEs has been the deciding factor to recognize these events as the signatures of newly reconnected tubes at the dayside magnetopause and then fastly moving poleward (Sandholt et al., 1992). One striking feature is the large potential drop inferred across these localized and short-lived structures: occasionally, it may reach values comparable to the large-scale convection potential drop across the magnetosphere ockwood et al., 1989). This raises fundamental questions on the nature of the reconnection processes: it is not yet understood whether they operate with bursty transient events such as these Flux Transfer Events, or with continuous phenomena as initially suggested, or with a mixture of both (Cowley and Lokwood, 1992). These questions address in turn the large-scale convection, precisely driven by the reconnection processes, its response to an impulsive or continuous excitation, and the resulting flow dynamics inside the magnetospheric cavity.

Substorms. The term "substorm" also designates impulsive and intense events. They take place in the nightside plasmasheet, and if the onset is probably very localized, their effects modify the global plasma distribution and the large-scale magnetic configuration over several tens of Earth's radii. Their signature in the nightside ionosphere, very early identified from all-sky cameras and ground-based magnetometer arrays, still remains relevant nowadays (Akasofu, 1964). After a quiet sequence of auroral arcs smoothly drifting equatorward in the nightside auroral oval, the sudden intensification of the equatorward arc near midnight indicates the substorm onset. A bulge of bright and active auroras then develops, it expands poleward and then both eastward and westward around the pole, while magnetometers detect an intense circulation of ionospheric currents: this is the expansion phase, which can last a few tens of minutes. A longer recovery phase follows: the auroral brightness is fading and the poleward edge of the auroral oval retreats equatorward.

These substorm signatures in the auroral ionosphere imply a triggering process in the nightside plasmasheet due to the strong coupling existing between both regions. Spacecraft observations at the geostationary orbit near the plasmasheet inner edge also reveal a characteristic sequence of effects in the midnight sector (Sauvaud and Winckler, 1980). Before the substorm onset, the magnetic field lines are progressively stretching tailward and the plasmasheet is simultaneously getting thinner: the plasmasheet currents flowing across the tail come closer to the Earth and intensify. Then occurs a brief and spectacular phase, called "dipolarization", when the magnetic configuration gets suddenly back to the initial dipolar configuration. At the geostationary orbit and in the midnight sector, it is accompanied by dispersionless injections of energetic particles, thus indicating the close location of an injection boundary or a simultaneous propagation with the dipolarization. Observations at larger distances in the tail report similar plasmasheet thinnings and frequently fast tailward flows. They provide evidences that substorm effects do not remain confined to the auroral oval or to the inner plasmasheet but deeply modify the whole magnetotail configuration.

Despite the number of observations accumulated since a few decades in various regions of the terrestrial environment, the fundamental questions concerning major magnetospheric events like substorms remain unsolved. The ultimate triggering processes are not definitely identified. It is often admitted that the solar wind might play an important role. Occasionally, correlations can been found with southward turnings of the interplanetary magnetic field, which provide favorable conditions to an increase of the reconnection rate at the dayside magnetopause. However substorms also occur in many other conditions. Moreover, the nature of the processes occurring at the onset, and even the location or the extension of the onset region, still remain controversial. Modeling efforts have explored various working assumptions (Lui, 1991). One of the most widespread models suggests the formation of an X-type reconnection point, or rather of an X-line, possibly in the near-Earth plasmasheet at an estimated distance between 10 and 30 Earth's radii: it would account for the plasmasheet thinning, for the earthward injection and auroral precipitation of energetic particles on one side of the neutral line and for the tailward expulsion of fast flows on the other side. Other models assume an eventual failure of the magnetospheric magnetic configuration to support the observed intensification of the cross-tail currents, thus causing the disruption of these currents and their deflection into the auroral ionosphere. Instabilities (tearing mode, ballooning mode, ...) are also mentioned as possible causes for substorm onset.

Although very different, FTEs and substorms share similar features. They are not only impulsive and transient events localized on the dayside or in the nightside plasmasheet. Because their effects propagate to the major part of the magnetosphere and deeply modify its magnetic configuration and the plasma transport, both appear as major magnetospheric events. Despite the numerous observations accumulated for years, fundamental questions remain unsolved. The nature of the triggering processes, the location, characteristics and dynamics of the onset region are not definitely identified. Different working assumptions presently address crucial problems, such as the conditions for magnetic reconnection or other mechanisms to develop at magnetospheric plasma boundaries, the relative contribution of bursty localized events and of quasi-steady large-scale effects, the response of the magnetospheric plasma and of the global convection transport to these processes.

6   Conclusions

The existence of the magnetospheric cavity around a planet depends on the dominant energy sources stored in the interacting bodies: the planet and the surrounding interplanetary medium. In the case of the Earth, the internal magnetic field sets a magnetic obstacle against the dynamic pressure exerted by the supersonic super-Alfvénic flow of solar wind. Upstream from the magnetosphere, the bow shock contributes to decelerate the solar wind at subsonic and sub-Alfvénic velocities such that the flow can be deflected around the obstacle. The magnetosphere's boundary, the magnetopause, prevents both the penetration of solar wind particles inside the cavity and the diffusion of the planetary field outside. Like a Faraday cage, the currents flowing at the magnetopause act as a shield for the terrestrial environment against the interplanetary medium.

The magnetopause location can be simply estimated in the framework of fluid theories from the equilibrium conditions at the magnetopause balancing the solar wind dynamic pressure and the magnetospheric magnetic pressure. The simulations of these conditions reproduce satisfactorily the observed magnetopause shape, compressed on the dayside and elongated to large distances on the nightside, the solar wind flow around the magnetospheric obstacle and the draping of interplanetary magnetic field lines around it.

Spacecraft observations inside the magnetosphere reveal the presence of plasmas, electric fields; in particular, two auroral ovals permanently encircling the poles exhibit variations correlated with solar activity. They suggest that the shielding effect of the magnetopause is imperfect, and that mass, momentum and energy transfers take place between the solar wind and the magnetosphere. This assumption does not contradict the first and successful approach based on an analysis of the dominant energy sources stored in the planet and in the interplanetary medium. It rather suggests a more comprehensive description including secondary processes in terms of energetics. In this regard, the magnetized feature of the solar wind plasma plays a central role. The interplanetary magnetic field fluctuates and its interaction with the terrestrial magnetic field may produce an X-type reconnection figure if they happen to be equal and opposite at some places. Such configurations are responsible for the solar plasma entry into the magnetosphere. This is the concept of open magnetosphere first proposed by Dungey (1961).

Secondary regarding the energetics, reconnection processes have huge implications for the magnetosphere. The first of them is certainly to provide a particle source of solar origin in addition of the ionosphere, which is the only internal plasma source in the terrestrial magnetosphere. This solar wind plasma entry is accompanied by a momentum transfer which excites a large-scale "convection" motion inside the magnetosphere: the plasma describes two vortices involving an antisunward flow above each pole closed by a return motion in the sunward direction in the internal regions of the magnetosphere. Driven by the convection motion, the solar particles that have penetrated into the magnetosphere have then access to different parts of the magnetosphere and contribute to populate them. For example, the boundary layers just inside the magnetopause, and thus at proximity of the entry regions, reveal density and energy characteristics very similar to the solar wind. In the plasmasheet extended around the equatorial plane, the particles have gained energy by adiabatic heating during their sunward transport in the inner part of the magnetosphere. Some of them collide with the upper atmospheric layers and excite neutral constituents which relax by emitting these spectacular auroras, visible in the high-latitude sky.

Finally, the model of reconnecting magnetosphere provides a successful guideline to explain the global plasma and field distributions observed in the magnetosphere and in the ionosphere. It is mainly based on the opening of the magnetosphere and on the existence of mass, momentum and energy transfers with the solar wind, which then excite a large-scale convection motion inside the magnetospheric cavity. Despite the number of observations accumulated in various regions of the magnetosphere, the involved transfer processes remain poorly understood. An impulsive picture is progressively adding to or substituting for the smooth and quasi-steady description. For example, repeated series of the so-called Flux Transfer Events discovered at the dayside magnetopause are interpreted as the signature of an impulsive and bursty reconnection, induced by the solar wind variability. Reconnection processes are also involved in substorms, these impulsive and bursty events that occur in the nightside magnetotail and deeply modify the configuration of a major part of the magnetosphere. Their various phases follow very characteristic sequences observed since a long time in the magnetosphere and in the ionosphere. However, for such fundamental phenomena, the crucial questions on the triggering factors, on the nature of the involved processes, even on their precise location are not yet elucidated. They are at the root of the motivations for a three-dimensional exploration of the magnetosphere and of its plasma boundaries as proposed by the CLUSTER mission (2000).

References

1. Akasofu S.-I. (1964) The development of an auroral substorm. Planet. Space Sci. **12**, 273
2. Axford W.I., Hines C. O. (1961) A unified theory of high-latitude geophysical phenomena and geomagnetic storms. Can. J. Phys. **39**, 1433
3. Baumjohann W., Treumann R. A. (1997) Basic Space plasma Physics. published by Imperial College Press
4. Chapman S., Ferraro V. C. A. (1931) A new theory of magntic storms. J. Geophys. Res. **36**, 79
5. Cowley S. W. H. (1982) The causes of the convection in the earth's magnetosphere: a review of the developments during the IMS. Rev. Geophys. **30**, 531
6. Cowley S.W.H., Lockwood M. (1992) Excitation and decay of solar-wind-driven flows in the magnetosphere-ionosphere system. Ann. Geophys. **10**, 103
7. Dungey, J. W. (1961) Interplanetary magnetic field and the auroral zones. Phys. Rev. Lett. **6**, 47
8. Harel M., Wolf R. A., Spiro R. W., ReiffP. H., Chen C. K. (1981) Quantitative simulations of a magnetospheric substorm, 1. Model logic and overview. J. Geophys. Res. **86**, 2217
9. Kennel C. F. (1995) Convection and Substorms. Paradigms of Magnetospheric Phenomenology, International Series on Astronomy and Astrophysics, edited by Dalgano et al., Oxford University Press
10. Kivelson M. G., Russell C. T. (1995) Introduction to Space Physics. edited by M.G. Kivelson and C. T. Russel, Cambridge University Press
11. Lockwood M., Wild M.N. (1993) On the quasiperodic nature of the magnetopause Flux transfer Events. J. geophys. Res. **98**, 5935
12. Lockwood M., Sandholt P.E., Cowley S.W.H. (1989) Dayside auroral activity and magnetic flux transfer from the solar wind. Geophys. Res. Letters **16**, 33
13. Lui A. T. (1991) Extended consideration of a synthesis model for magnetospheric substorms. in Magnetospheric Substorms, Geophys. Monogr. Ser. Vol. 64, edited by Kan et al., AGU, Washington, D.C.,pp 43
14. Ogino T., Walker R. J., Ashour-Abdalla M., Dawson J. M. (1986) An MHD simultion of the effects of the interplanetary magnetic field B_y component on the

interaction of the solar wind with the earth's magnetosphere during southward IMF. J. Geophys. Res. **91**, 10029

15. Onsager T.G., Fuselier S.A. (1994) The location of magnetopause reconnection for northward and southward interplanetary magnetic field, Solar system plasmas in space and time, Geophysical Monograph 84, AGU, pp 183

16. Peymirat C., Fontaine D. (1994) Numerical simulation of magnetospheric convection including the effect of field-aligned currents and electron precipitation. J. Geophys. Res. **99**, 11155

17. Russell C.T., R.C. Elphic R.C. (1979) ISEE observations of Flux Transfer Events at the dayside magnetopause. Geophysical Res. Lett. **6**, 33

18. Sandholt P.E., Locwood M., Denid W.E., Elphic R.C., Leontjev S. (1992) Dynamical auroral structure in the vicinity of the polar cusp: multipoint observations during southward and northward IMF. Ann. Geophys. **10**, 483

19. Sauvaud J. A., Winckler J. R. (1980) Dynamics of plasma, energetic particles, and fields near synchronous orbit in the nighttime sector during magnetic substorms. J. Geophys. Res. **85**, 2043

20. Southwood D. J.(1987) The ionospheric signature of flux transfer events. J. Geophys. Res. **92**, 3207

21. Spreiter J. R., Summers A. L., Alksne A. Y. (1966) Hydrodynamic flow around the magnetosphere. Planet. Space Sci. **14**, 223

22. Stern D. P. (1977) Large-scale electric fields in the Earth's magnetosphere. Rev. Geophys. Space Phys. **15**, 156

Self-Consistent Kinetic Approach for Low Frequency and Quasi-static Electromagnetic Perturbations in Magnetic-Mirror Confined Plasmas

René Pellat[1], Olivier Le Contel[2], Alain Roux[2], Sylvaine Perraut[2], Omar Hurricane[3], and Ferdinand V. Coroniti[4]

[1] Centre de Physique Théorique, Ecole Polytechnique, F–91128 Palaiseau Cedex, France
[2] Centre des Environnements Terrestre et Planétaires, CNRS–Université de Versailles/St Quentin, 10–12, avenue de l'Europe, F–78140 Vélizy, France
[3] Lawrence Livermore National Lab, P.O. Box 808, L–312, Livermore, CA 94550, USA
[4] Department of Physics, University of California at Los Angeles, Los Angeles, California 90095-1547 USA

Abstract. We describe a new self-consistent kinetic approach of collisionless plasmas. The basic equations are obtained from a linearization of the cyclotron and bounce averaged Vlasov and Maxwell equations. In the low frequency limit the Gauss equation is shown to be equivalent to the Quasi-Neutrality Condition (QNC). First we describe the work of Hurricane *et al.*, 1995b, who investigated the effect of stochasticity on the stability of ballooning modes. An expression for the energy principle is obtained in the stochastic case, with comparisons with the adiabatic case. Notably, we show how the non adiabaticity of ions allows to recover a MHD-like theory with a modification of the polytropic index, for waves with frequencies smaller than the bounce frequency of protons. The stochasticity of protons can be due, in the far plasma sheet (beyond 10-12 R_E, R_E being the Earth radius), to the development of thin Current Sheet (CS) with a curvature radius that becomes smaller than the ion Larmor radius. Conversely the near Earth plasma sheet (6-8 R_E), where the curvature radius is larger, is expected to be in the adiabatic regime. We give a description of slowly evolving (quasi-static) magnetic configurations, during the formation of high altitudes CS's, for instance during substorm growth phase in the Earth magnetosphere, and tentatively during the formation of CS's in the solar corona. Thanks to the use of a simple equilibrium magnetic field, a 2D dipole, the linear electromagnetic perturbations are computed analytically as functions of a forcing electrical current. The QNC, which is valid for long perpendicular wavelength electromagnetic perturbations ($k_\perp \lambda_D \ll 1$ where λ_D is the Debye length), is developed via an expansion in the small parameter T_e/T_i. To the lowest order in T_e/T_i ($T_e/T_i \to 0$) we find that the enforcement of the QNC implies the presence of an electrostatic potential which is constant along the field line, but varies across it. The corresponding potential electric field is perpendicular to the magnetic field; it corresponds to the self-consistent response of the plasma to an externally applied time varying perturbation. This potential electric field tends to reduce the effect of the induced electric field, hence producing a partial "shielding" of the motion that would correspond to the induced electric field if it was alone. The effect of the total azimuthal electric field, obtained from the QNC, on the radial transport of the plasma is investigated. We show that the direction of the perpendicular electric field varies with the latitude. As a consequence, for a time dependent transport, the equatorial electric field cannot usually be mapped onto the low altitude electric field (ionosphere for the Earth), even in the absence of a parallel electric field. Present calculations show that during the substorm growth phase, the (total) azimuthal electric field is directed eastward, close to the equator, and westward off-equator. Thus, large equatorial pitch-angle particles drift tailward whereas small pitch-angle particles drift earthward. Finally, to the next order in T_e/T_i, we show that the formation of the thin current sheet lead to the development of a finite parallel electric field. Thus time variations in high altitude CS's are coupled to the low altitude regions (ionosphere for the Earth) via (i) an electrostatic component constant along the magnetic field line and via (ii) the parallel electric fields. Associated with this parallel electric field, a parallel current develops. We suggest that this current drives an instability at frequencies well above that imposed by the forcing current. Unstable waves are electromagnetic and have frequencies of the order of the proton gyrofrequency. Given their large amplitudes these waves can produce a fast electron and ion diffusion which modify the electrical currents in a such manner that the reconfiguration of the magnetic field occurs.

1 Introduction

Sudden releases of large amounts of magnetic energy, presumably due to plasma instabilities, occur in many different contexts: solar flares, magnetospheric substorms, "saw-tooth" events (tokamak). In all cases, the plasma confinement is lost; over a short time interval particles (electrons and protons) are heated/accelerated, and powerful non thermal radiations are emitted. Before the rapid release of energy, a quasi-static stage is observed as evidenced by the formation of a thin Current Sheet (CS). Since it gives a self-consistent global picture of plasma dynamics, one can be tempted to use the Magneto-HydroDynamics (MHD) to describe this energy release. If the characteristic time between binary collisions is much shorter than the time variations associated with the energy release, the regime is collisional wich ensures that MHD is a valid description. Conversely, in a collisionless regime the conditions of validity of MHD are quite restrictive, in particular for an inhomogeneous plasma, as we will see later on. For the Earth's plasma sheet, where substorms take place, the plasma is always collisionless; the mean-free path for binary collisions between particles is larger than the size of the magnetosphere. In Tokamaks the fast dynamical events, the so-called "saw-tooth" events, that correspond to a decrease in the current, and to a loss of confinement, occur over a time scale too short to be controled by collisions. In the solar corona it is difficult to estimate the typical size of active regions and to compare it with the mean free path between binary collisions. In a collisionless plasma, ideal MHD does not allow for magnetic reconfiguration, thus some form of departure from ideal MHD is often assumed to take place in a so called diffusion region, thereby allowing cross-B diffusion and initiating magnetic reconnection. Several authors assume the existence of an "anomalous" parallel resistivity to introduce a relation between the parallel current and the parallel electric field, as if there were a resistivity that would allow for the departure from ideal MHD. The concept of an "anomalous" parallel resistivity was developed by Sagdeev and Galeev who suggested that it could result from the emission, by electrons, of plasma waves driven unstable by a field aligned current, followed by the absorption, by ions, of these waves [1]. While interesting as a concept, this idea has not been confirmed by numerical simulations; various attempts made to give evidence for a stationary resistivity along the field lines failed [2]. Therefore one should not invoke an anomalous resistivity associated with waves-particle interactions to justify the use of non ideal MHD in a collisionless regime. There are also other difficulties, associated with the geometry, that restrict the validity of MHD. In a mirror geometry like magnetic field lines in the solar corona or in planetary magnetospheres, during non-stationary events such as solar flares or substorms, there are important reasons why the usual MHD approach may fail. Classically, the use of MHD equations (Ohm's law) implies the following restricting conditions on the frequency and on the wave length of the perturbations: $\omega < \Omega$ and $k_\perp \rho < 1$ (Ω being the gyrofrequency, $k_\perp$ the perpendicular component of the wave vector and ρ the Larmor radius). Moreover, the system of momentum equations deduced from Liouville's equation is a priori infinite, thus one has to close it with an equation of state, which is equivalent to fixing a well defined relation between the LOCAL values of the parameters. For instance one often assumes that the divergence of the heat flux is null, which corresponds to the classical adiabatic approximation (thermodynamical sense). The equation of state being a local approximation, it is not valid when there are resonances, which introduce non-local effects. In the case of an homogeneous equilibrium, the description of electromagnetic perturbations by MHD equations, with an adiabatic closure equation, are valid, as long as there are very few resonant particles, which is achieved as soon as the resonant velocity is much larger than the thermal velocity: $v_{res} \gg v_{th}$ (the thermal velocity). This can be rewritten: $\omega \gg k_\parallel v_{th}$ for Landau resonance (or $\omega - n\Omega_i \gg k_\parallel v_{th}$ for cyclotron resonances, $k_\parallel$ being the parallel component of the wave vector and Ω_i being the ion gyrofrequency). This condition ensures that there are few resonant particles; one can therefore neglect Landau (and cyclotron) damping. It gives a low limit for the frequency of the studied perturbation In an inhomogeneous medium, however, there are other limitations, associated for instance with field line curvature, and the associated bounce motion. See also discussions on the importance of these resonances in [3], [4],[5] and [6]. For instance, as soon as the parallel wave length of the pertubations is equivalent to $L_\parallel$, the length of the magnetic field line, the above condition writes $\omega \gg v_{th}/L_\parallel \simeq \omega_b$ (ω_b being the particle bounce frequency). Then, the bounce frequency of electrons (and ions) is a low limit for the use of MHD equations and the full conditions of validity write : $\omega_b < \omega < \Omega_i$.

Thus, against a widely spread wisdom, MHD is not a priori valid to describe the low frequency perturbations ($\omega < \omega_b$) associated with the magnetic field reconfiguration in a collisionless plasma. In view of what has just been said, the ratio between the collision frequency and the electron bounce time can be used as a test to know whether the regime is collisional or collisionless. Thus in order to find whether magnetic reconfiguration is controled by collisions or is occuring in a collisionless regime, we have compared the electron-electron collision frequency with the electron bounce frequency through CS's. Estimates of the electron density versus altitude, in the solar corona, are given in a companion paper by Demoulin and Klein [7]. Adopting the values given in the figure 4 and the table 2 of Demoulin and Klein, and assuming a temperature of 100 eV at and beyond 1.1 R_S (R_S being the solar radius), we concur with their conclusion, and find that the low altitude Loops (R ≥ 1.1 R_S, N $\simeq 10^9$ to 10^{10} cm^{-3}) are dominated by collisions, because the electron-electron collision frequency (the largest of the collision frequencies) is: $\nu_{ee} \simeq 50$ to 500 s^{-1}, which is much larger than the electron bounce frequency: $f_{be} \simeq 4 \times 10^{-2}$ s^{-1}. Conversely for the Streamers: R $\geq$ 3-10 R_S, N $\simeq 5 \times 10^5$ to 5×10^6 cm^{-3}, the collision frequency: $\nu_{ee} \simeq 0.025$ to 0.25 s^{-1}, is of the order of the electron bounce frequency through the CS: $f_{be} \simeq 0.04$ s^{-1}, for 100 eV and for a CS thickness of 10^8 m. Thus electron bounce can occur at least inside the CS and has to be taken into account to write the QNC as described below (Sect. 3), irrespective of collisions. For Sheets (R $\simeq$ 1.5 to 2 R_S), the density is 10^7 to 10^8 cm^{-3}, at an altitude between 1.5 and 2 R_S. With these parameters the collision frequency seems to be still larger than the bounce frequency of electrons. It is obvious, however, that some strong pressure gradient/pressure anisotropy is needed to producee these thin currents sheets, and that the electron thermal energy is enhanced (few keV, few tens of keV?) inside the CS. Observations show that before the energy dissipation occurs, the CS gets thinner and thinner (see [8] and references therein). It is difficult to understand how an increasingly thin CS can be produced in a collisional regime. Similarly the existence of a non-thermal radiation generated in these regions proves that collisionless processes are important in determining the dynamics of active solar events. In the rest of the paper we will consider the collisionless regime which certainly applies to the Earth's plasma sheet and seems to apply to Streamers and possibly to Sheets, but not to Loops.

From the observational point of view, it is easier to estimate the rate of energy dissipation occuring during solar flares or substorms than the electron-electron collision frequency; therefore another way (independently of the estimation of the electron-electron collisions) to check if the regime is collisional or collisionless is to compare this rate of energy dissipation with the electron bounce period. For the Earth's magnetosphere, at 7 R_E (R_E = Earth radius),

τ_b is typically 1 sec for energetic electrons (1 keV) and 50 sec for ions (10 keV), this latter time is precisely the characteristic time of dissipation. With the values quoted above, taken from Demoulin and Klein (this issue), we get, for Streamers and Sheets: $T_{be} \simeq 25$ s and $T_{bi} \simeq 15$ mn. The solar events occur on a time scales of 10^2 - 10^3 s, comparable to T_{bi}, but apparently larger than the collision time. The hard X-ray-spikes and the radio-spikes, however, occur over much shorter time scale; 10^{-1} s, and 10^{-2} s, respectively, comparable to the electron bounce time through the CS. Then electron dynamics is probably not controled by collisions. The case of ions is a little more difficult. Indeed, even if the ion bounce period is shorter than the characteristic time of dissipation, the ion dynamics may be "collision like" due to the stochasticity, which occurs when the curvature radius is equivalent to the ion Larmor radius. In this case, a strong pitch-angle scatterring occurs for each crossing of the magnetic equatorial plane and plays the role of "pseudo-collisions" [9]. The pitch-angle scattering may be also produced by wave interaction with "high-frequency" ion-cyclotron waves that can eventually be present in the medium. For resonant ion-cyclotron waves with an amplitude of the order of 1 % of the static magnetic field, as observed in situ at magnetospheric substorms ($\delta B/B \simeq 10^{-2}$), a resonant velocity of the order of the Alfvén velocity, and a spectral width $\Delta\omega \simeq \omega_{H+}$ (ω_{H+} the proton gyropulsation), one gets a diffusion time of the order of 1 to 10 msec, shorter than the collision time. However only in situ measurements will be able to provide an estimate of the amplitude of resonant ion-cyclotron waves. The effect of pitch-angle diffusion on the reconfiguration of the magnetic field will be discussed later on.

The second most important limitation of MHD is the implicit assumption that the electric field component parallel to the magnetic field lines is null. During substorms, there are some strong, though indirect evidences, for field-aligned potential drops. In particular large potential drops have been found above the auroral ionosphere within an altitude range of 5000-15000 km, causing the energization of the electrons responsible for the aurorae. In this region, the parallel electric field may be large. In the case of the Earth's magnetosphere, an indirect evidence for parallel electric fields is provided by electrons beams observed near the magnetic equator during and after substorm breakup [10, 11]. We are not aware of direct/indirect evidence for parallel E-fields, in the solar corona, though X-ray emissions imply the existence of accelerated/highly non-thermal electrons [12].

Thus we conclude from the above discussion that in the Earth plasma sheet, the plasma is undoubtely collisionless during all phases of the substorm, and that MHD is not valid for the description of substorm growth-phase (formation of a CS) and breakup (magnetic reconfiguration). Indeed, the caracteristic times of these phases are longer than the electron (and ion) bounce frequency. Regarding reconnection in the solar corona it seems difficult to give a definitive answer without carying out in situ measurements.

The previous discussion suggests that Loops are likely to be in a collisional regime, and can therefore be described by MHD. The magnetic pressure being much larger than the kinetic, the regime is force free, for Loops. Conversely at higher altitudes (beyond 1.5 R_S) the pressure gradient and and/or thermal anisotropy should be large enough to withstand thin CS's. Inside these CS's the temperature is higher and the electron bounce period across the CS an/or ion resonant diffusion is/are likely to occur over time shorter than collisions. Thus MHD is not a priori applicable; a kinetic theory is needed.

The first attempt to describe substorm as the result of the spontaneous development of a kinetic instability was made, longtime ago, by Coppi *et al.* (1966) who suggested that collisionless tearing modes can develop in the geomagnetic tail, for a simple geometry with no magnetic field across the sheet [13]. Later Lembege and Pellat (1982) have shown that even a very small perpendicular magnetic field component stabilizes the tearing modes. More generally the electron finite compressibility has been shown to prevent the development of tearing modes [14]. Thus another collisionless process has to be identified. The purpose of the present paper is to present a new approach of collisionless magnetic reconfiguration based upon the study of low frequency ($\omega < \omega_b$) electromagnetic perturbations in the direction of the equilibrium current.

2   The Bounce Averaged Solution of the Linearized Vlasov Equation

In this section, we give the gyrokinetic solution to the Vlasov equation, which allows for the development of perturbations with $k_\perp \rho \simeq 1$ ($k_\perp$ being the wave number across the field and ρ the particle Larmor radius) [15]. Assuming that the electromagnetic perturbation is periodic in time and in space (across to the field line) we take:

$$\delta\Phi(r,t), \delta A(r,t) = \widehat{\delta\Phi}(k_\perp,\omega,l), \widehat{\delta A}(k_\perp,\omega,l) \exp\left(i(k_\perp \cdot r_\perp + \omega t)\right),$$

where r (resp. $k_\perp$) is the position vector (resp. the wave vector) and $\perp$ denotes the component perpendicular to the magnetic field. For the sake of simplicity we omit the $\widehat{}$ symbol and the exponential factor in the following formulas, then the linearized response δf is given by:

$$\delta f = q\frac{\partial f_0}{\partial E}\left(\delta\Phi - u_y\delta A_y + \left(1 + \frac{\omega_\star}{\omega}\right)\lambda e^{-iS} - \left(1 + \frac{\omega_\star}{\omega}\right)g\right) \tag{1}$$

where $f_0(E, p_y)$ is the equilibrium distribution function (E is the particle energy and p_y the canonical momentum), u_y is the diamagnetic drift velocity, $\omega_\star = k_y u_y$ is the diamagnetic drift frequency (for more details about $\omega_\star$ see for instance [16]), k_y is the wave number in the y-direction (azimuthal). We work in local field-aligned coordinates defined by the triad of unit vectors:

$$e_\chi = \frac{B}{B}, \; e_\psi = \frac{\nabla\psi}{|\nabla\psi|} \text{ and } e_y = e_\chi \times e_\psi. \quad (2)$$

In this frame, the velocity becomes:

$$v = |v_\perp|(e_\psi \cos\xi + e_y \sin\xi) + v_\parallel e_\chi. \quad (3)$$

To obtain the linear response (1), a change of variables $(v_\psi, v_y, v_\parallel) \rightarrow (E, \mu, \xi)$ has been made, where $E = \frac{1}{2}mv_\parallel^2 + \mu B$ is the kinetic energy, $\mu = 1/2mv_\perp^2/B$ is the magnetic moment and ξ is the gyrophase angle. The elementary volume in velocity space becomes $d^3v = \sum_{\sigma=-1,+1} BdEd\mu d\xi/(m^2|v_\parallel|)$ where $\sigma = \text{sign}(v_\parallel)$ (for more details see [17, 18, 19, 20]). In (1), the function g contains the non local wave-particle interaction. To the first order in ω/ω_b, the function g becomes

$$g = e^{-iS}\left(\frac{\omega}{\omega + \overline{\omega_d}}\overline{H} + \sigma i\omega \int \frac{dl}{|v_\parallel|}\left[H - \frac{\omega + \omega_d}{\omega + \overline{\omega_d}}\overline{H}\right]\right), \quad (4)$$

$S = k_\perp v_\perp \sin(\alpha_k - \xi)/\Omega$, $\alpha_k = \text{Arctan}(k_\psi/k_y)$, Ω is the cyclotron frequency, $\omega_d = k_y v_d$ is the gradient-curvature drift frequency, the upper bar denotes bounce averaging and H is given by:

$$H = J_0\left(\delta\Phi - v_d\delta A_y\right) + \frac{(\omega + \omega_d)}{\omega}\lambda + iJ_1 v_\perp\left(\frac{k_\psi}{k_\perp}\delta A_y - \frac{k_y}{k_\perp}\delta A_\psi\right), \quad (5)$$

where $\lambda = i\omega \int^l dl' J_0\delta A_\parallel$ and J_n are Bessel functions of argument $k_\perp|v_\perp|/\Omega$. In the next section we substitute the linearized solution of the Vlasov equation into the QNC. The bounce averaged linear solution of Vlasov equation obtained here is similar to those developed and used by different authors e.g. [21, 5, 4].

3   The Quasi-neutrality Condition

In this section, we develop the QNC via an expansion in the small parameter T_e/T_i. The validity of this expansion for the Earth's magnetotail is suggested by several observations indicating that this ratio is small in the magnetotail. For instance, a statistical study of current disruptions from AMPTE/CCE when the spacecraft was in the near-Earth current sheet is presented in [22]. The authors showed that the electron to proton temperature ratio is in the range of 0.11 to 0.57. They pointed out that these values are higher than those reported in [23] based on IRM data. Indeed, data presented in [23], yield average plasma properties, notably an electron to proton temperature ratio in the range: 0.09-0.18. These authors also noticed that this ratio is nearly the same as the one found in [24] at distances of $|X|$=30-60 R$_E$. More recently, during a dusk-dawn crossing of the near-Earth tail by GEOTAIL, during a relatively quiet period, a ratio around 0.2 was measured [25]. In any cases, it is therefore possible to consider T_e/T_i as a small parameter over a wide range of radial distances from the Earth, and for different levels of activity.

We do not know whether this assumption is valid for solar flare conditions but it is possible to develop the QNC without this assumption [20].

From the linear response of the plasma (equations 1 to 5) and assuming that f_0 is a maxwellian distribution function ($f_0 = n_0 \left(m/(2\pi T)\right)^{3/2} \times \exp -(E/T)$), the QNC: $\sum_{j=i,e} q_j \int d^3v \, \delta f_j \simeq 0$ can be written:

$$\sum_{j=i,e} (\frac{q_j}{m_j})^2 \frac{1}{T_j} \int \frac{4\pi B dE d\mu}{|v_\parallel|} f_{0j} \times \left[\delta\Phi + \left(\frac{\omega + \omega_\star}{\omega}\lambda - \frac{\omega + \omega_\star}{\omega + \overline{\omega}_d}\overline{H}\right)J_0\right]_j \simeq 0$$

$$(6)$$

where we have performed the gyrophase integration and summed over streaming and anti-streaming velocities. This latter operation cancels out the part of δf that is an odd function of σ. The above relation was derived earlier in [20], where more details are given about the derivation. The gauge $\delta A_y = 0$ has been chosen. for the sake of simplicity, the usual wave length ordering $k_\perp \rho_j \ll 1$ is made. In this limit, the Bessel functions become $J_0 \simeq 1$ and $J_1 \simeq k_\perp |v_\perp|/2\Omega$, and the expression for H simplifies; we get:

$$H = \delta\Phi + \lambda + \Xi \quad \text{where} \quad \Xi = \frac{\omega_d \lambda}{\omega} - i\frac{\mu k_y \delta A_\psi}{q}. \tag{7}$$

Then, after some algebraic manipulations, the QNC can be written as

$$\sum_{j=i,e} (\frac{q_j}{m_j})^2 \frac{1}{T_j} \int \frac{4\pi B dE d\mu}{|v_\parallel|} f_{0j} \left[(\delta\Phi - \overline{\delta\Phi}) + (\lambda - \overline{\lambda})\right] =$$

$$- \sum_{j=i,e} (\frac{q_j}{m_j})^2 \frac{1}{T_j} \int \frac{4\pi B dE d\mu}{|v_\parallel|} f_{0j} \left[\frac{\overline{\omega}_{dj} - \omega_{\star j}}{\omega + \overline{\omega}_{dj}}\left(\overline{\delta\Phi} + \overline{\lambda}\right)\right.$$

$$\left. - \frac{\omega + \omega_{\star j}}{\omega + \overline{\omega}_{dj}}\left(\overline{\Xi}_j\right)\right], \quad (8)$$

where the terms $\omega_{\star j}\lambda$ cancel between electrons and ions, because $u_{yi}/T_i + u_{ye}/T_e = 0$, [20]. Note that the diamagnetic drift frequency and the purely magnetic drift frequency of electrons can easily be related to the corresponding terms for ions: $\overline{\omega}_{de}^{th}(\alpha) = -T_e/T_i \, \overline{\omega}_{di}^{th}(\alpha)$ and $\omega_{\star e} = -T_e/T_i \, \omega_{\star i}$ (where th means thermal quantities and α denotes the particle pitch-angle). The QNC becomes :

$$\int \frac{4\pi BdEd\mu}{m_e^2|v_\parallel|} f_{0e} \left[(\delta\Phi - \overline{\delta\Phi}) + (\lambda - \overline{\lambda}) + \frac{T_e}{T_i}\left(\delta\Phi - \overline{\delta\Phi}) + (\lambda - \overline{\lambda}) \right) \right] =$$

$$\frac{T_e}{T_i} \int \frac{4\pi BdEd\mu}{m_i^2|v_\parallel|} f_{0i} \left[\frac{\overline{\omega}_{di}\,(\overline{\omega}_{di} - \omega_{\star i})}{\omega(\omega + \overline{\omega}_{di})} (\overline{\delta\Phi} + \overline{\lambda}) \right.$$

$$\left. + \frac{\omega_{\star i} - \overline{\omega}_{di}}{\omega + \overline{\omega}_{di}} (\overline{\Xi}_i) \right]. \quad (9)$$

Since the RHS term of (9) is proportional to (T_e/T_i), we get to the lowest order in (T_e/T_i):

$$\int \frac{4\pi BdEd\mu}{m_e^2|v_\parallel|} f_{0e} \left[(\delta\Phi - \overline{\delta\Phi}) + (\lambda - \overline{\lambda}) \right] = 0 + O(T_e/T_i). \quad (10)$$

A trivial solution of (10) is:

$$\delta\Phi + \lambda = \Phi_0(\psi,\, y) + O(T_e/T_i), \quad (11)$$

where Φ_0 is constant for a given magnetic field line. This constant component of the perturbed electrostatic potential is always taken to be equal to zero ($\delta\Phi + \lambda = 0$) in studies based on MHD. An external electrostatic field, modeling the convection, is often added to the inductive part of the electric field in order to better fit the data. In the present paper, we show that the quasi-neutrality over the volume of the flux tube implies that Φ_0 is different from zero. We compute Φ_0 in a self-consistent manner, as a function of the electromagnetic perturbation defined by λ and $\delta B_\parallel$. Integrating the QNC (9) over the volume of the flux tube, we find

$$\int \frac{dl}{B} \left[\int \frac{4\pi BdEd\mu}{m_i^2|v_\parallel|} f_{0i} \times \left[\frac{\overline{\omega}_{di}\,(\overline{\omega}_{di} - \omega_{\star i})}{\omega(\omega + \overline{\omega}_{di})} (\Phi_0) + \frac{\omega_{\star i} - \overline{\omega}_{di}}{\omega + \overline{\omega}_{di}} (\overline{\Xi}_i) \right] \right] = 0,$$
$$(12)$$

where the LHS term of (9) has vanished thanks to the identity $\int dl/B \int d^3v$ $(X - \overline{X}) = 0$, valid for any function $X(E,\, \mu,\, l)$. Finally, Φ_0 writes

$$\Phi_0 = \frac{\int \frac{dl}{B} \left[\int \frac{4\pi BdEd\mu}{m_i^2|v_\parallel|} f_{0i} \times \left[\frac{\omega_{\star i} - \overline{\omega}_{di}}{\omega + \overline{\omega}_{di}} (\overline{\Xi}_i) \right] \right]}{\int \frac{dl}{B} \left[\int \frac{4\pi BdEd\mu}{m_i^2|v_\parallel|} f_{0i} \times \left[\frac{\overline{\omega}_{di}(\omega_{\star i} - \overline{\omega}_{di})}{\omega(\omega + \overline{\omega}_{di})} \right] \right]}. \quad (13)$$

Now, we can calculate the self-consistent perturbed electric field. Taking into account the implications of (11) to the lowest order in T_e/T_i, the perturbed electric field $\delta \boldsymbol{E} = -\boldsymbol{\nabla}(\delta \Phi) - \partial/\partial t(\delta \boldsymbol{A})$, in the y-direction, becomes (remembering that $\delta A_y = 0$):

$$\delta E_y = -ik_y(\delta \Phi) = -ik_y(\Phi_0 - \lambda). \tag{14}$$

Thus, the full perpendicular electric field, associated with the perturbation, is the sum of an inductive component (λ) plus an electrostatic component (Φ_0) determined from the QNC. This electric field will produce a transport of the plasma. Notice that this transport is different from a steady convection; it is associated with an electromagnetic perturbation (see discussion in introduction). The electrostatic component, associated with Φ_0, tends to reduce the effect of the inductive component of the electric field λ, thereby producing a partial shielding of the motion that would correspond to the inductive electric field (if it was not shielded).

The expression $\lambda = i\omega \int^l dl' \delta A_\parallel$ shows that the partial derivative of λ with respect to l is equal to the inductive component of the parallel electric field ($\partial \lambda/\partial l = \partial \delta A_\parallel/\partial t$). Thus, locally and in the limit $T_e < T_i$, (11) implies that the inductive component of the parallel electric field $\partial \delta A_\parallel/\partial t$ is balanced by the parallel gradient of the perturbed electrostatic potential $\partial \delta \Phi/\partial l$. Hence, to the lowest order in $T_e/T_i < 1$, (11) is equivalent to the usual MHD approximation, where one assumes the absence of a parallel electric field ($E_\parallel = -\partial/\partial l(\delta \Phi + \lambda) = 0$). In the present study, the absence of a parallel electric field is not an assumption but an (approximate) result, obtained by solving the QNC in the limit $T_e < T_i$. We show, however, in Sect. 5.3, that a finite parallel electric field exists to the order T_e/T_i and give some important consequences of this parallel electric field.

Finally, the potential Φ_0 affects the stability of the plasma as was demonstrated in the simple electrostatic multipole case in [26]. In the next section, we explore the effects of Φ_0 on the stability, in a fully electromagnetic case.

4   Plasma Stability

4.1   Adiabatic Regime

With the expressions (1) to (5) for the linearized response of the plasma, one can obtain a kinetic variational form by combining δf with Ampère's law and the QNC. The original variational form obtained in [15] does not take into account the existence of Φ_0; it writes

$$\int dy d\psi \frac{dl}{B} \frac{|\nabla \times \delta \mathbf{A}|^2}{\mu_0} = 2\pi \sum_{j=i,e} \left(\frac{q}{m}\right)^2 \int dy d\psi dE d\mu \oint \frac{dl}{|v_{\parallel}|} \left(\frac{f_0}{T}\right)_j$$

$$\times \left\{ |\phi|^2 + \frac{\omega + \omega_\star}{\omega} \left[|\lambda + I|^2 - |I|^2 \right] \right.$$

$$\left. + \frac{(\omega + \omega_\star)\omega_d}{\omega^2} |\lambda|^2 - \frac{\omega + \omega_\star}{\omega + \overline{\omega}_d} |\overline{H}|^2 \right\}_j , \quad (15)$$

where $I = H - \lambda(\omega + \omega_d)/\omega$ and H given by (5). For details about the derivation of (15) see [18]. With the wavelength ordering $k_\perp \rho_j \ll 1$ and including Φ_0, the variational form becomes

$$\int dy d\psi \frac{dl}{B} \frac{|\nabla \times \delta \mathbf{A}|^2}{\mu_0} = 2\pi \sum_{j=i,e} \left(\frac{q}{m}\right)^2 \int dy d\psi dE d\mu \oint \frac{dl}{|v_{\parallel}|} \left(\frac{f_0}{T}\right)_j$$

$$\times \left\{ |\Phi_0|^2 + (\omega + \omega_\star)\frac{X}{k_y}\frac{2\mu B}{q}\left(Y + \frac{\partial X}{\partial \psi}\right) \right.$$

$$\left. + (\omega + \omega_\star)\omega_d \left|\frac{X}{k_y}\right|^2 - \frac{\omega + \omega_\star}{\omega + \overline{\omega}_d} |\overline{H}|^2 \right\}_j , \quad (16)$$

where cross terms like XY are to be read as $\text{Re}(XY^\star)$ and Re stands for the real part, $X = k_y \lambda/\omega$, $Y + \partial X/\partial \psi = ik_y \delta A_\psi/B$. The X and Y variables are similar to those defined in [27] with $\nabla \times \delta \mathbf{A} = \nabla \times (\boldsymbol{\xi} \times \mathbf{B})$ and defining $X = B\xi_\psi$ et $Y = ik_y\xi_y$ (see also[19, 20]). The vector $\boldsymbol{\xi}$ is the usual MHD displacement vector of the plasma from his equilibrium position see [27]. From (11), the expression of H is now $H = \Phi_0 + \Xi$ where $\Xi = (\omega_d/\omega)\lambda - i\mu k_y A_\psi/q$. To obtain (16), we used the following expressions for the local gradient-curvature drift frequency and diamagnetic drift frequency respectively $\omega_d = -(k_y v_{\parallel} m/JBq)\partial(JBv_{\parallel})/\partial\psi$ and $\omega_\star = k_y p'/qn_0$ [19, 20].

Performing the velocity space integration and taking into account the expression (13) of Φ_0, we obtain

$$\delta W = \int dy d\psi \frac{dl}{B} \left[\frac{1}{\mu_0}\left[\left(\frac{\partial X}{\partial l}\right)^2 + \left(\frac{B}{k_y}\frac{\partial Y}{\partial l}\right)^2 + B^2\left(Y + \frac{\partial X}{\partial \psi}\right)^2\right] \right.$$

$$+ 2p'X\left(Y + \frac{\partial X}{\partial \psi}\right) + p'X^2 \frac{\partial \ln J}{\partial \psi}\Big]$$

$$+ 2\pi \left(\frac{q}{m}\right)^2 \int dy d\psi dE d\mu \oint \frac{dl}{|v_{\parallel}|} \left(\frac{f_0}{T}\right)_i$$

$$\times \left[\frac{\overline{\omega}_{di}(\overline{\omega}_{di} - \omega_{\star i})}{\omega(\omega + \overline{\omega}_{di})} |\Phi_0|^2 - \frac{\omega + \omega_{\star i}}{\omega + \overline{\omega}_{di}} |\overline{\Xi}_i|^2 \right], \quad (17)$$

One should notice that the expression of δW is virtually identical to the δW of Bernstein *et al.* (equation 6.16) with the following exceptions:

- there is no Z variable (the parallel fluid displacement) due to the bounce averaging that takes place in the plasma response function,
- the compressibility term (the last term between brackets) is computing by taking into account the existence of the Φ_0.

4.2 Stochastic Regime

As mentioned in the introduction, when the local radius of curvature is of the order of the thermal ion Larmor radius, the response of the ions is modified. A stochastic ion experiences an apparently random pitch angle scattering on each crossing midplane. This scattering is followed by adiabatic motion to the bounce point and back to midplane. Hurricane *et al.* found that inclusion of stochastic ion dynamics replaces the bounce averages in the gyrokinetic solution with flux tube volume averages [18, 28]:

$$\overline{Q} = \frac{\int_0^{l_b} \frac{dl}{v_\parallel} Q}{\int_0^{l_b} \frac{dl}{v_\parallel}} \rightarrow \langle Q \rangle = \frac{\int_0^{E/B_n} d\mu \int_0^{l_b} \frac{dl}{v_\parallel} Q}{\int_0^{E/B_n} d\mu \int_0^{l_b} \frac{dl}{v_\parallel}}, \tag{18}$$

where Q is any quantity, l_b is the bounce point of an ion for a given value of magnetic moment μ, B_n is the value of the magnetic field at the bounce point. In this treatment, μ is a stochastic variable distributed with an uniform probability; all quantities are which is averaged upon. In this regime, it is possible to carry out the calculations and to compute explicitly Φ_0 and δW.

With the frequency ordering $\omega > \omega_\star, \overline{\omega}_d$ and in the context of stochastic dynamics, (13) becomes to the lowest (nonvanishing) order

$$\Phi_0 = \omega \frac{\int \frac{dl}{B} \left[\int \frac{4\pi B dE d\mu}{m_i^2 |v_\parallel|} f_{0i} (\omega_{\star i} - \langle \omega_{di} \rangle) \langle \Xi \rangle \right]}{\int \frac{dl}{B} \left[\int \frac{4\pi B dE d\mu}{m_i^2 |v_\parallel|} f_{0i} (\omega_{\star i} \langle \omega_{di} \rangle) - (\langle \omega_{di} \rangle^2) \right]}. \tag{19}$$

After the velocity space integration (19) gives [20]

$$\Phi_0 = \frac{\omega}{k_y} C \frac{v'}{v''}, \tag{20}$$

where C is the flux tube average compressibility defined by

$$C = \frac{\oint dl \nabla \cdot \boldsymbol{\xi}/B}{\oint dl/B} = \frac{\int dl/B \, (X \partial \ln J/\partial \psi + Y + \partial X/\partial \psi)}{\int dl/B}, \tag{21}$$

and

$$\frac{v'}{v''} = \frac{\partial \left(\ln \oint \frac{dl}{B} \right)}{\partial \psi}. \tag{22}$$

Using (20) and changing the bounce averages of adiabatic theory by stochastic averages defined by (18), the quadratic form becomes in the limit $k_y \to \infty$ but with $k_\perp \rho_i \ll 1$ [20]:

$$\delta W_s = \int dy d\psi \left[\oint \frac{dl}{B} \left(\frac{\partial X}{\partial l} \right)^2 + p' \left(\oint \frac{dl}{B} X^2 D - \frac{(\oint \frac{dl}{B} X D)^2}{\oint \frac{dl}{B} D} \right) \right],$$

(23)

where the subscript "s" denotes "stochastic" and $D = \partial ln J / \partial \psi - \mu_0 p' / B^2$ is proportional to the magnetic field line curvature. By the Schwarz inequality, the term in the large parentheses is positive if D is positive everywhere, and negative if D is negative everywhere. In the Earth's magnetotail, for instance, the pressure gradient is directed toward the Earth ($p' < 0$) and the curvature is likely to be convex ($D > 0$) so that the second term is destabilizing and could lead to $\delta W_s < 0$, if the stabilizing contribution from $\partial X / \partial l$ was small enough. Furthermore, one should notice that the inclusion of Φ_0 modifies the polytropic index of the plasma since the results without Φ_0 are equivalent to results with Φ_0 when γp is changed by $-p'v'/v''$ (where γ is the polytropic index) [20].

In this stochastic regime, the particles pressure is isotropized by the pitch angle scattering associated with the loss of adiabaticity. Finally, it can be shown that the solution of the QNC, in the stochastic case (stochastic ions and electrons), gives no parallel electric field. Indeed, in this case, the RHS of equation (9) vanishes exactly. Thus, waves that develop in the magnetotail (23) at distance large enough for ions and electrons to be non adiabatic can be called "MHD–like" waves [18, 20].

5   Plasma Transport During the Substorm Growth Phase

In the Earth's magnetotail, one regularly observes a slow change in the magnetic configuration: the magnetic field lines are stretched corresponding to the formation of a thin CS; this change will be modelled as a quasi-static perturbation. Therefore, in this section, we use the linear self-consistent kinetic approach developed in the previous sections to study the transport of the plasma in the NEPS, in response to quasi-static variations of the magnetic field. Thus, the theory developed for electromagnetic perturbations with a time scale longer than the bounce period of electrons and ions ($\omega \ll \omega_{be}, \omega_{bi}$), and for spatial scales larger than the ion Larmor radius ($k_\perp \rho \ll 1$) is applied. Furthermore, we assume that $\omega \ll k_\parallel v_A$, namely that the time scale of the growth phase is large compared with typical travel time of Alvfén waves. Due to this quasi-static assumption, a particular treatment of the Ampere's law has been done. Unlike substorm injection which is known to be a sudden process ($\omega \geq \omega_{bi}$) with a small spatial scale ($k_y \to \infty$), the build-up of a tail-like configuration is a slow process ($\simeq 30$ min) affecting a large fraction of the tail (small k_y). Thus the applied electromagnetic perturbation (during growth phase) is not considered as being the consequence of a local internal instability, as it is probably the case for breakup. The change from a dipole to a tail-like configuration, instead, is considered as the result of the response of the magnetotail to a quasi-static forcing caused by variations in the solar wind (e.g. [29]). The full treatment of this problem is very difficult since it would require a full description of the forcing caused by the solar wind, taking into account the boundary conditions imposed at the magnetopause. Moreover, the way the solar wind drives the stretching of the magnetic field lines is still not completely understood. To simplify, we assume that the change of the dipolar field close to the Earth is due to an increase of the current further out in the tail (e.g. [29]). Thus, to the lowest order in β (where β is the ratio between the kinetic pressure and the magnetic pressure) the Ampere's law gives $\nabla \times \boldsymbol{B} = 0$. To the next order, the linearization gives $\nabla \times \delta \boldsymbol{B} = \mu_0 \delta \boldsymbol{j}_{ext}$ where $\delta \boldsymbol{j}_{ext}$ is a perturbed current located far from the dipolar region. In the following subsection, we solve the first order of the Ampere's law and give the components of the perturbed electromagnetic field as a function of an external forcing via an electrical current.

5.1   Ampère's Law

Close to the Earth, we can neglect the local electrical currents which corresponds to assuming $\beta < 1$. Thus, we can approximate the field by a dipole. To allow us to carry out analytical calculations, we use a two-dimensional (2D) dipole [30] to describe the equilibrium magnetic field. Using cylindrical coordinates (r, θ, y) where θ is the colatitude, the 2D magnetic field model is defined by:

$$\boldsymbol{B} = -\frac{\hat{D}}{r^2} \left(\cos \theta \boldsymbol{u_r} + \sin \theta \boldsymbol{u_\theta} \right), \tag{24}$$

where $\hat{D}$ is the dipolar moment. The magnetic field strength is given by

$$B = \frac{B_{eq}}{\sin^2 \theta}, \tag{25}$$

where $B_{eq} = \hat{D}/L^2$, is the equatorial magnetic field strength, L is the equatorial-crossing distance of the relevant field line. For the 2D dipole, the local coordinates become:

$$\psi = -\hat{D}/L, \ y \text{ and } \chi = \hat{D} \cot \theta / L. \tag{26}$$

It follows that $\boldsymbol{B} = \nabla \psi \times \boldsymbol{e_y}$ and the magnetic field strength is

$$B = \frac{\psi^2}{\hat{D} \sin^2 \left(\text{arccot} \left(-\chi/\psi \right) \right)}. \tag{27}$$

The bounce period and the bounce average curvature-gradient magnetic drift velocity are

$$\tau_b = \frac{2\pi L}{v} \text{ and } \overline{v_d} = \oint \frac{dl}{v_\parallel} v_d = \frac{-2E}{qLB_{eq}} e_y. \tag{28}$$

Then, we perturb the dipolar equilibrium by an external current, flowing in the westward direction and located far in the tail. The linearized Ampère's law becomes

$$\nabla \times \delta \boldsymbol{B} = \mu_0 \delta j_{ext} \boldsymbol{e}_y \tag{29}$$

We assume that

$$\delta \boldsymbol{B}(\boldsymbol{r}, t) = \widehat{\delta \boldsymbol{B}}((\psi, k_y, l, \omega) \exp{(i(k_y y + \omega t))},$$

for the sake of simplicity we omit the ^symbol and the exponential factor in the following formulas. The external current δj_{ext} is defined by

$$\delta j_{ext}(L, k_y, \theta, \omega) = \delta j_{eq}(k_y, \omega)\delta(L - L_c)\sin^{2n}\theta\left((2n+1)\cot^2\theta - 1\right). \tag{30}$$

We have assumed that the current is highly localized in radial distance and we choose, for simplicity, a Dirac function $\delta(L - L_c)$ where L_c is the radial location of the forcing current. Therefore, we are interested in L values between $0 < L < L_c$ where the 2D dipole assumption is valid. Along the field line, we have chosen a class of forcing current with a dependence that allows us to obtain easily the magnetic field perturbation. This forcing current must also correspond to an increase in the equatorial current as suggested by the observations (e.g. [31, 32]). This class is labelled by an index n, the larger n, the more localized is the perturbation close to the magnetic equator (see Fig. 1). Comparison between results obtained for various n gives insight on how sensitive the results are to the θ dependence of the forcing current. For $n = 0$, the perturbed current is divergent at high latitudes but as we will check later on, this divergence does not modify the results because the perturbed components of the electromagnetic field do not diverge. For the sake of simplicity, in the course of the paper, we often use the case $n = 0$ to obtain estimates of the various characteristic quantities.

After some algebra (described in Appendix A) and, in the limit $|k_y|L > 1$ and $|k_y(L - L_c)| < 1$, the perturbed components of the magnetic field write:

$$\delta B_\psi = -\frac{\mu_0 |k_y| \delta j_{eq}(k_y, \omega)}{2} L_c^2 \left(\sin^{2n+1}\theta\cos\theta\right), \tag{31}$$

$$\delta B_\parallel = -\frac{\mu_0 \delta j_{eq}(k_y, \omega)}{2} \frac{L_c^2}{L} \sin^{2n}\theta\left((2n+1) - (2n+2)\sin^2\theta\right)$$
$$\times \left(H(L - L_c) - H(-(L - L_c))\right), \tag{32}$$

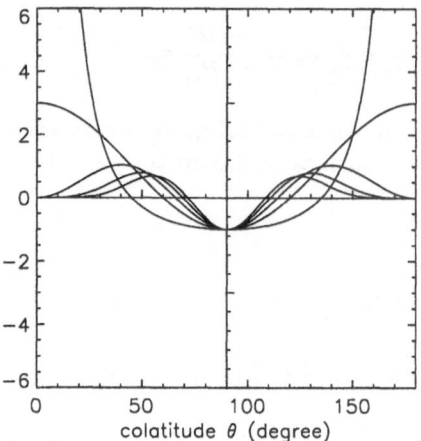

Fig. 1. Variation of the external current $\delta j_{yext}(\theta) \propto \sin^{2n}\theta\left((2n+1)\cot^2\theta - 1\right)$ versus the colatitude θ for $n = 0, 1, 2, 3, 4$.

where H is the Heaviside function. For λ, we obtain (Appendix A):

$$\lambda_n(L, k_y, \theta, \omega) = \lambda_{neq}\left(c_n(L, k_y) + (\sin^2\theta)^{n+1}\right),\tag{33}$$

where we have defined $c_n(L, k_y) = (n+1)!/(-|k_y(L-L_c)|)^{n+1}$ and $\lambda_{neq} = 1/(4(n+1))\mu_0(|k_y|/k_y)\omega\delta j_{eq}(k_y, \omega)L_c^2 L$.

To obtain, the real components of the perturbed magnetic field, we have to perform an inverse Fourier transform in time and in y. We find (see Appendix B):

$$\delta B_\psi = \frac{\mu_0}{2}L_c^2\left(\sin^{2n+1}\theta\cos\theta\right)P\left(\int_{-\infty}^{\infty}dy'\frac{\delta j_{eq}(y', t)}{(y-y')^2}\right),\tag{34}$$

$$\delta B_\| = -\frac{\mu_0}{2}\frac{L_c^2}{L}\sin^{2n}\theta\left((2n+1) - (2n+2)\sin^2\theta\right)\delta j_{eq}(y, t)\tag{35}$$

$$\times\left(H(L-L_c) - H(-(L-L_c))\right),\tag{36}$$

where P denotes the Cauchy principal value. Now, we have to specify the variation of the forcing current in the y-direction. During the growth phase, spacecraft observations close to midnight [33, 31, 34] show that the magnetic field changes from a dipole-like configuration to a tail-like configuration. The equatorial value of the magnetic field decreases whereas, off-equator, the radial component increases. The duration of this variation is typically about 30-45 minutes. Moreover, breakup is usually observed to start close to midnight in a longitudinally narrow sector while the rest of the magnetotail keeps on stretching [35]. These observations suggest that while the reconfiguration at breakup is localized in longitude, the formation of the current sheet during the growth phase is more homogeneous in longitude. Thus, while the limit $k_y \to \infty$ is adapted to study breakup, the formation of the current sheet can be better described by a finite k_y. Therefore, we consider an external current localized around the noon-midnight meridian, flowing eastward and slowly increasing with the time as

$$\delta j_{eq}(y,t) = \delta j_m \exp(-\frac{y^2}{\Delta^2}) \exp(\gamma t), \tag{37}$$

where δj_m is the initial magnitude of the current, $1/\gamma$ is the characteristic time scale of the growth phase and Δ is the characteristic scale along y where the tail current increases. The complete expression for the external current becomes:

$$\delta j_{ext}(L,y,\theta,t) = \delta j_{eq}(y,t)\delta(L - L_c) \sin^{2n}\theta \left[(2n+1)\cot^2\theta - 1\right], \tag{38}$$

We verify that for $\theta = \pi/2$ (magnetic equator), the forcing current flows westward as suggested by observations. Then, we can compute the ψ component of the perturbed magnetic field which gives:

$$\delta B_\psi = -\frac{\mu_0 \delta j_m}{\sqrt{\pi}} L_c^2 \left(\sin^{2n+1}\theta \cos\theta\right) \frac{1}{\Delta} \left[1 + \zeta P\left(\tilde{W}(\zeta)\right)\right] \tag{39}$$

where $\tilde{W}(\zeta) = 1/(\sqrt{\pi}) \int_{-\infty}^{\infty} dV \exp(-V^2)/(V-\zeta)$, is the Fried-Conte function and we have defined $V = y'/\Delta$ and $\zeta = y/\Delta$. Close to midnight, $\zeta < 1$, and in this limit, the Fried-Conte function can be approximated by $P(\tilde{W}(\zeta)) \simeq -2\zeta + O(\zeta^3)$ and we obtain:

$$\delta B_\psi = -\frac{\mu_0 \delta j_m}{\sqrt{\pi}} L_c^2 \left(\sin^{2n+1}\theta \cos\theta\right) \frac{1}{\Delta} \left(1 - 2\left(\frac{y}{\Delta}\right)^2\right). \tag{40}$$

One should notice that in the opposite limit $\zeta > 1$ (far away of the maximum of the current in the y direction), the expansion of the Fried-Conte function is $-1/\zeta$ and $\delta B_\psi = 0$. Finally, close to midnight ($\zeta < 1$), the two perturbed components of the magnetic field write:

$$\delta B_\psi = -\frac{\mu_0 \delta j_m}{\sqrt{\pi}} L_c^2 \left(\sin^{2n+1}\theta \cos\theta\right) \frac{1}{\Delta} \left(1 + 2\left(\frac{y}{\Delta}\right)^2\right) \tag{41}$$

$$\delta B_\parallel = -\frac{\mu_0}{2} \frac{L_c^2}{L} \sin^{2n}\theta \left((2n+1) - (2n+2)\sin^2\theta\right) \delta j_{eq}(y,t)$$
$$\times (H(L - L_c) - H(-(L - L_c))) \tag{42}$$

We verify that for a forcing current directed westward ($\delta j_m > 0$) at the magnetic equator ($\theta = \pi/2$), the radial component of the equilibrium magnetic field increases off-equator whereas the parallel component near the equator decreases, which corresponds to observations carried out during the growth phase (see Fig. 2).

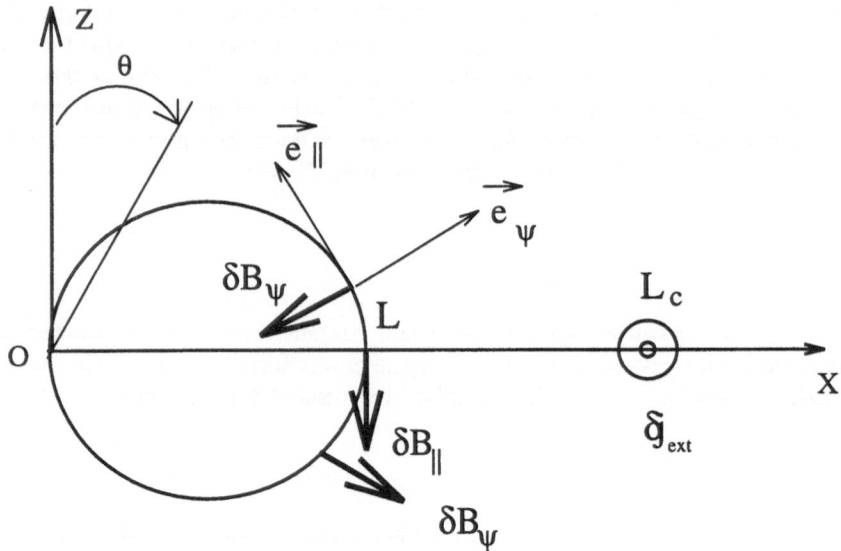

Fig. 2. Schematic diagram of the electromagnetic perturbation applied on a 2D dipole field to model the change of configuration which occurs during the growth phase. As indicated by the arrows, the magnetic field perturbation tends to produce a tail-like configuration

5.2   Transport of the Plasma

In the previous Sects. 2, 3, and Sect. 5.1, we have completely solved the Vlasov-Maxwell system of equations in the quasi-static limit ($\omega < k_\parallel v_A$ and $\omega < \omega_b$). In (5) only the terms $\overline{\omega_d \lambda}/\omega$ and $\overline{i\mu k_y \delta A_\psi}/q = -\overline{\delta B_\parallel}/q$ appear. It is useful to compare the size of these two terms. Reminding that $\overline{\omega_d \lambda}/\omega \simeq k_y \overline{v_d} \lambda_{neq}/\omega$ and $\overline{\mu \delta B_\parallel}/q = -\mu B\overline{\overline{Y}}/q \simeq -E\lambda_{neq}/\omega/\hat{D}/q \simeq 1/(k_y L)k_y \overline{v_d}\lambda_{neq}/\omega$, we can conclude that in the limit $|k_y|L > 1$ we have $\overline{\mu \delta B_\parallel}/q < \overline{\omega_d \lambda}/\omega$. Therefore the term containing $\delta B_\parallel$ can be neglected we take

$$\overline{\mu \delta B_\parallel}/q \simeq 0. \tag{43}$$

In this case, the linearized Vlasov equation, which describes the behavior of the plasma, can be simplified, (1) and (4) remain the same but the expression (5) of H becomes:

$$H = \left(J_0 \delta\Phi + \frac{(\omega + \omega_d)}{\omega}\lambda\right), \tag{44}$$

where $\delta\Phi$ is given by the QNC (9) that implies $\delta\Phi = \Phi_0 - \lambda$, with Φ_0 given by (13), and λ and $\delta B_\parallel$ given by (33) and (42) (from the Ampère's law solved in the limit $|k_y|L > 1$ and $|k_y(L - L_c)| < 1$). Now, to obtain the self-consistent perpendicular electric field associated with the magnetic field perturbations, we need to compute the constant part, Φ_0, of the perturbed electrostatic potential. Taking into account that $\overline{\mu \delta B_\parallel}/q \ll \overline{\omega_d \lambda}/\omega$, the expression, (13), of Φ_0 becomes:

$$\Phi_0 = \frac{\int \frac{dl}{B} \left[\int \frac{4\pi B dE d\mu}{m_i^2 |v_\parallel|} f_{0i} \times \left[\frac{\omega_{*i} - \overline{\omega}_{di}}{\omega + \overline{\omega}_{di}} \left(\overline{\left(\frac{\overline{\omega}_{di} \lambda}{\omega}\right)} \right) \right] \right]}{\int \frac{dl}{B} \left[\int \frac{4\pi B dE d\mu}{m_i^2 |v_\parallel|} f_{0i} \times \left[\frac{\overline{\omega}_{di}(\omega_{*i} - \overline{\omega}_{di})}{\omega(\omega + \overline{\omega}_{di})} \right] \right]}.$$ (45)

The expression of Φ_0 (see Appendix D) becomes

$$\Phi_0 = (c_n + S_n) \lambda_{neq},$$ (46)

where we have defined

$$S_n = \sum_{k=0}^{n+1} (-1)^k C_{n+1}^k \frac{(2k-1)!!}{(2k+2)!!} \sum_{j=0}^{k} (-1)^j C_k^j \frac{(2j)!!}{2^{j-1}(j+1)!}$$
$$\times \left(1 + k \frac{(2j+2)(j+3/2)}{(2j+3)(j+2)} \right). \quad (47)$$

Now, from (14), the self-consistent perpendicular electric field writes:

$$\delta E_y = -ik_y \lambda_{neq} \left(S_n - (\sin^2 \theta)^{n+1} \right).$$ (48)

After an inverse Fourier transform, we obtain:

$$\delta E_y = \frac{\mu_0 L_c^2 L}{4(n+1)} \frac{1}{\pi} \left(S_n - (\sin^2 \theta)^{n+1} \right) \frac{\partial}{\partial t} \left(P \left(\int_{-\infty}^{\infty} dy' \frac{\delta j_{eq}(y', t)}{(y-y')^2} \right) \right).$$ (49)

Taking into account the expression of the current (37), we find

$$\delta E_y = -\frac{\mu_0 \delta j_m \gamma L_c^2 L}{2(n+1)} \frac{1}{\sqrt{\pi}} \left(S_n - (\sin^2 \theta)^{n+1} \right)$$
$$\times \frac{1}{\Delta} \left(1 + \zeta P \left(\tilde{W}(\zeta) \right) \right) \exp(\gamma t). \quad (50)$$

Then, in the limit $\zeta < 1$, we obtain

$$\delta E_y = \delta E_{L,y,t}(L, y, t) \frac{1}{n+1} \left(S_n - (\sin^2 \theta)^{n+1} \right),$$ (51)

where we have defined :

$$\delta E_{L,y,t}(L, y, t) = -\frac{\mu_0 \delta j_m \gamma L_c^2 L}{2} \frac{1}{\sqrt{\pi}} \frac{1}{\Delta} \left(1 - 2 \left(\frac{y}{\Delta} \right)^2 \right) \exp(\gamma t).$$ (52)

The colatitude θ where the perpendicular electric field changes sign is given by:

$$\theta_0 = \arcsin\left(S_n^{1/(2n+2)}\right). \tag{53}$$

Because S_n is always smaller than unity, the direction of the perpendicular electric field changes along the field line even in the absence of a parallel electric field (see Fig. 3). As noted in Sect. 3, δE_y is directed eastward (positive) close to the equator, it is null for $\theta = \theta_0$, and it is directed westward (negative) for $\theta < \theta_0$. The larger n, the larger is θ_0 therefore the region where δE_y is eastward gets thinner (as n increases). As an example, for $n = 0$, $S_0 = 5/6$, $\lambda_0 = \left(c_0 + \sin^2 \theta\right)\lambda_{0eq}$, $\Phi_0 = \left(c_0 + 5/6\right)\lambda_{0eq}$ and $\delta E y = \delta E_m(L, y, t)\left(5/6 - \sin^2 \theta\right)$. δE_y is directed eastward (positive) close to the equator, it is null for $\theta = \arcsin(5/6)^{1/2}$ and is directed westward (negative) for $\theta < \arcsin(5/6)^{1/2}$. From (52), we can estimate the intensity

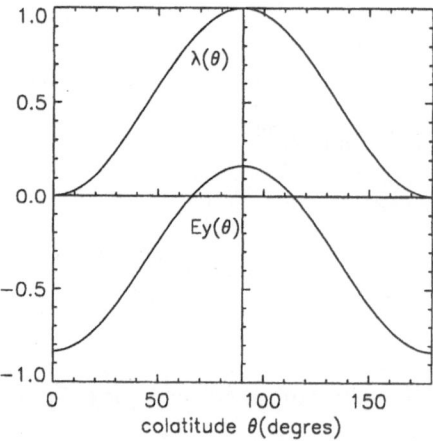

Fig. 3. Variation of λ and δE_y versus the colatitude θ for n=0.

of the perpendicular electric field during the growth phase. For instance, the characteristic variation of the equatorial magnetic field at the geostationary orbit ($L_c \simeq L = 6.6\ \mathrm{R}_E$) is of order of 30 nT for a duration of the growth phase of 30 minutes. We assume that the radial scale of the current sheet is about 1 R_E therefore $\mu_0 \delta j_m \simeq 30 \times 10^{-9}/(6.4 \times 10^6)$T/m. In the y direction, we assume that the spatial scale Δ is about 4 R_E. We obtain for the inductive component of the perpendicular electric field (without the contribution of Φ_0), $\delta E_{L,y,t} \simeq 2.5\ \mathrm{mV/m}$ which is reduced to 0.4 mV/m at the equator due to the contribution of the electrostatic component Φ_0 (S_n in (51)). A smaller

Δ would result in a larger value of $\delta E_{L,y,t}$. As we have previously mentioned, the effect of the Φ_0 is to decrease the magnitude of the total electric field compared to the inductive component. Therefore, the plasma transport is also reduced.

Next, we can compute the bounce average electric drift to study the motion of the particles along ψ as a function of their pitch angle. We obtain (see Appendix E):

$$\overline{u_{E_y}} = \frac{1}{\tau_b} \oint \frac{dl}{|v_\parallel|} \frac{\delta E_y}{B} = \frac{\delta E_{L,y,t}(L,y,t)}{B_{eq}} \frac{1}{n+1} \left(\frac{S_n}{2} \left(1 + \frac{\mu B_{eq}}{E} \right) \right.$$
$$\left. - \sum_{k=0}^{n+2} (-1)^k C_{n+2}^k \frac{(2k-1)!!}{(2k)!!} \sum_{j=0}^{n+2} (-1)^j C_k^j \left(\frac{\mu B_{eq}}{E} \right)^j \right). \quad (54)$$

We note that the bounce average electric drift, associated with δE_y, depends on the magnetic moment. To simplify, we can consider two extreme cases: equatorial pitch-angle particles of $90°$ ($\mu B_{eq}/E \simeq 1$) and equatorial pitch-angle particles of $0°$ ($\mu B_{eq}/E \simeq 0$). We obtain:

$$\overline{u_{Ey}} \simeq \begin{cases} \delta E_{L,y,t}/B_{eq} \left(S_n/2 - \sum_{k=0}^{n+2}(-1)^k C_{n+2}^k (2k-1)!!/(2k)!! \right) \\ \text{for } \alpha_{eq} = 0°, \\ \delta E_{L,y,t}/B_{eq} (S_n - 1) \text{ for } \alpha_{eq} = 90°, \end{cases} \quad (55)$$

where $\delta E_{L,y,t}$, given by (52) is always negative close to the midnight meridian ($y < \Delta$). Since S_n is always smaller than unity the bounce average electric drift of $90°$ particles and that of $0°$ particles have opposite directions. $90°$ particles drift tailward whereas $0°$ particles drift earthward, during the magnetic field line stretching. When n increases, the perturbation is more and more localized close to the equator and S_n decreases. Therefore, from (55), we deduce that $90°$ particles drift more and more tailward whereas $0°$ particles drift slowly eartward. Again for $n = 0$, we find:

$$\overline{u_{Ey}} \simeq \begin{cases} 1/24(\delta E_{L,y,t}(L,y,t)/B_{eq}) \text{ for } \alpha_{eq} = 0°, \\ -1/6(\delta E_{L,y,t}(L,y,t)/B_{eq}) = -4u_{Ey}(0°) \text{ for } \alpha_{eq} = 90°. \end{cases}$$

Thus, the correct treatment of the QNC implies a perpendicular motion in response to a quasi-static electromagnetic perturbation ($\omega < k_\parallel v_A$ and $\omega < \omega_b$). Yet, because the perpendicular electric field direction varies with the position along the field line, the bounce averaged motion is different for different pitch-angles. Ninety degrees pitch-angle particles, which mirror close to the equator drift tailward while zero degree particles drift earthward. These results are different from the results obtained by a steady convection approach where the convection across a static magnetic field is ensured by an imposed electrostatic field (e.g. [30]). In the present work, the transport is due to the self-consistent response of the plasma to the quasi-static perturbation including the necessity of enforcing the quasi-neutrality.

5.3   Self-consistent Parallel Electric Field

To determine the field-aligned potential drop which can exist to the first order in (T_e/T_i), we have to solve the QNC to this order. Thus, in this case: $\delta\Phi(l) + \lambda(l) = \Phi_0 + \widetilde{\delta\phi}(l)$, where $\widetilde{\delta\phi}(l)$ is a first order term proportional to (T_e/T_i). Equation (9) becomes:

$$\int \frac{4\pi B dE d\mu}{m_e^2 |v_\parallel|} f_{0e} \left(\widetilde{\delta\phi} - \overline{\widetilde{\delta\phi}} \right) = \frac{T_e}{T_i} \int \frac{4\pi B dE d\mu}{m_i^2 |v_\parallel|} f_{0i}$$

$$\times \left[\frac{\overline{\omega}_{di} \left(\overline{\omega}_{di} - \omega_{*i} \right)}{\omega(\omega + \overline{\omega}_{di})} \Phi_0 + \frac{\omega_{*i} - \overline{\omega}_{di}}{\omega + \overline{\omega}_{di}} \left(\Xi_i \right) \right]. \quad (56)$$

Provided that $\omega < \omega_b$, (56) is valid for any quasi-static or low frequency perturbations. Thus a parallel electric field will be present in response to electromagnetic perturbations applied in a mirror geometry like the near-Earth magnetic tail. Therefore, parallel electric fields should exist not only during active periods (breakup) or at low altitudes (inverted V), but also during relatively quiet periods such as the substorm growth phase, as will be seen in the next sections. Unfortunately, it is not possible to find a general analytical solution of the integral equation (56) except when ions and electrons are stochastic. In such a case, it can be shown that the RHS of (56) cancels out exactly due to fast pitch-angle scattering on the two species. Therefore, there is no parallel electric field for a completely stochastic plasma (stochastic ions and electrons).

It will be shown in the next subsection that an analytical solution can be obtained for the region close to the Earth considered in Sects. 5.1 and 5.2. In this region, electrons and ions are in an adiabatic regime and the β of the plasma can be assumed to be small therefore the local currents can be neglected. We show that a self-consistent parallel electric field, developing in response to the quasi-static change of the magnetic field lines, can be calculated in the near Earth plasma sheet.

General Case

Equation (56) defines the solution to the order T_e/T_i of the QNC, in response to a perturbation defined by (33), (42) and (46). Taking into account (165) and (43), M, the RHS term of (56), becomes:

$$M = \frac{T_e}{T_i} \int \frac{4\pi B dE d\mu}{m_i^2 |v_\parallel|} f_{0i} \left[\frac{\overline{\omega}_{di} \left(\overline{\omega}_{di} - \omega_{*i} \right)}{\omega \left(\omega + \overline{\omega}_{di} \right)} \right]$$

$$\times \left[S_n - 2 \sum_{k=0}^{n+1} (-1)^k C_{n+1}^k \frac{(2k-1)!!}{(2k+2)!!} \left(1 - \frac{\mu B_{eq}}{E} \right)^k \left(1 + k \frac{\mu B_{eq}}{E} \right) \right] \lambda_{neq}.$$

$$(57)$$

Remembering that for a 2D dipole, the bounce averaged curvature-gradient magnetic drift frequency does not depend upon the magnetic moment μ, the first term between brackets of the expression of M in (57) is also independent of μ. Therefore, the μ integration in (57) can easily be performed thanks to the relation (B.2). We obtain:

$$M = n_{0i} K_i \left(S_n - 2R_n\right) \lambda_{neq},\tag{58}$$

where K_i is defined by:

$$K_i = \frac{T_e}{T_i} \frac{1}{n_{0i}} \int \frac{4\pi\sqrt{2m_i E}dE}{m_i^2} f_{0i} \left[\frac{\overline{\omega}_{di}\left(\overline{\omega}_{di} - \omega_{*i}\right)}{\omega\left(\omega + \overline{\omega}_{di}\right)}\right],\tag{59}$$

and R_n by:

$$R_n = \sum_{k=0}^{n+1}(-1)^k C_{n+1}^k \frac{(2k-1)!!}{(2k+2)!!} \sum_{j=0}^{k}(-1)^j C_k^j \frac{(2j)!!}{(2j+1)!!} \left(\frac{B_{eq}}{B}\right)^j$$
$$\times \left(1 + k\left(\frac{B_{eq}}{B}\right)\frac{2j+2}{2j+3}\right).\tag{60}$$

In Sect. 5.2, we have seen that the transport depends very little on n. The only change introduced by modifing n is the degree of localization, close to the equator. For a larger n, the perturbation is localized closer to the magnetic equator and therefore the properties of the transport also change more rapidly along the field line. For simplicity, in the following calculations, we fix the value of n to zero. We consider a Fourier decomposition of $\widetilde{\delta\phi}$ with respect to the colatitude θ:

$$\widetilde{\delta\phi} = \sum_{p=1}^{\infty}\left(a_p \cos p\theta + b_p \sin p\theta\right).\tag{61}$$

The term $p = 0$ is not included because the structure of the LHS of (56), $(X - \overline{X})$, leads to the cancellation of any constant term. Thus, the first non-zero term begins for $p = 1$. The RHS term, M, can be written as a function of $\cos p\theta$; for $n = 0$, it is simply given by:

$$M = n_{0i} K_i \left[\frac{1}{30} + \frac{1}{15}\cos 2\theta - \frac{1}{60}\cos 4\theta\right]\lambda_{0eq}.\tag{62}$$

We notice that (i) the RHS of (62) contains only even powers of $\cos p\theta$, and (ii) the highest value of p in (62) is 4; higher harmonics vanish. We therefore assume that $\widetilde{\delta\phi}$ only contains terms in $\cos 2\theta$ and $\cos 4\theta$; an assumption that will be verified later, therefore the solution obtained from this expansion will be shown to be exact. With this expansion, the field-aligned perturbed potential $\widetilde{\delta\phi}$ writes:

$$\widetilde{\delta\phi} = \sum_{l=1}^{2} a_{2l} \cos 2l\theta. \tag{63}$$

The bounce average value of $\widetilde{\delta\phi}$ is:

$$\overline{\widetilde{\delta\phi}} = -\frac{\mu B_{eq}}{E}\left(a_2 + 2a_4\left(1 - \frac{3}{2}\frac{\mu B_{eq}}{E}\right)\right) \tag{64}$$

The remaining double integral, in the LHS of (56), can be calculated from (C.2) and by taking into account the following identity:

$$\int_0^\infty E^{k-1/2}e^{-E/T}dE = \frac{\sqrt{\pi}}{2^k}(2k-1)!!\,T^{k+1/2}, \ (k \geq 1). \tag{65}$$

Combining (63), (64) and (65), one gets:

$$\int \frac{4\pi BdEd\mu}{m_e^2|v_\parallel|} f_{0e}\overline{\widetilde{\delta\phi}} = -\frac{2}{3}n_{0e}\frac{B_{eq}}{B}\left[a_2 + 2a_4 - \frac{12}{5}a_4\frac{B_{eq}}{B}\right]. \tag{66}$$

With the help of the simple relation $B_{eq}/B = \sin^2\theta = (1 - \cos 2\theta)/2$, the expression (66) becomes:

$$\int \frac{4\pi BdEd\mu}{m_e^2|v_\parallel|} f_{0e}\overline{\widetilde{\delta\phi}} = -n_{0e}\left[\frac{1}{3}a_2 + \frac{1}{15}a_4\right.$$
$$\left. + \cos 2\theta\left(-\frac{1}{3}a_2 + \frac{2}{15}a_4\right) + \cos 4\theta\left(-\frac{1}{5}a_4\right)\right]. \tag{67}$$

We notice that the first two even terms in $\cos p\theta$, considered in the Fourier series, do not provide terms in $\sin p\theta$, nor terms in $\cos p\theta$ with p odd, nor terms in $\cos p\theta$ with $p \geq 4$. Thus, the higher order terms ($p > 4$), which have not been considered in the above expansion, are solution of the following system of equations:

$$\sum_p \xi_p X_p = 0, \tag{68}$$

where $X_p = b_p, a_{2l+1}, a_{2l}$. A trivial solution is:

$$b_p = 0 \ (\forall p), \ a_{2l+1} = 0 \ (\forall l), \ \text{and} \ a_{2l} = 0 \ (\text{for } l > 2). \tag{69}$$

It was therefore justified to neglect terms with $p > 4$ as well as terms in $\sin p\theta$ and $\cos p\theta$ (p being odd). Therefore, (62) and (67) allow us to write (56), as a function of $\cos 2\theta$ and $\cos 4\theta$, in the following form:

$$n_{0e}\left[a_2\cos 2\theta + a_4\cos 4\theta + \left\{\frac{1}{3}a_2 + \frac{1}{15}a_4\right.\right.$$
$$+ \cos 2\theta\left(-\frac{1}{3}a_2 + \frac{2}{15}a_4\right) + \cos 4\theta\left(-\frac{1}{5}a_4\right)\right\}\right] = n_{0i}K_i$$
$$\times\left[\frac{1}{30} + \cos 2\theta\frac{1}{15} + \cos 4\theta\left(-\frac{1}{60}\right)\right]. \quad (70)$$

Taking $n_{0e} = n_{0i}$ at the equilibrium, (70), which should be satisfied for any value of θ, is equivalent to a system of three equations:

$$\frac{1}{3}a_2 + \frac{1}{15}a_4 = \frac{1}{30}K_i, \quad (71)$$

$$\frac{2}{3}a_2 + \frac{2}{15}a_4 = \frac{1}{15}K_i, \quad (72)$$

$$\frac{4}{5}a_4 = -\frac{1}{60}K_i.. \quad (73)$$

Finally, from (72) and (73), we find immediately: $a_4 = -K_i/48$ and $a_2 = 5K_i/48$. Equation (71) is then identically verified. The perturbed electrostatic potential, obtained from the QNC to the first order in (T_e/T_i), is therefore:

$$\widetilde{\delta\phi} = K_i e(\theta), \quad (74)$$

where we have defined the function $e(\theta) = A_2\cos 2\theta + A_4\cos 4\theta$, with $A_2 = 5/48$ and $A_4 = -1/60$. The parallel electric field is given by:

$$\delta E_\| = -\frac{\partial}{\partial l}(\delta\Phi) - \frac{\partial}{\partial t}(\delta A_\|) \quad (75)$$

$$= -\frac{\partial}{\partial l}(\Phi_0 - \lambda + \widetilde{\delta\phi}) - \frac{\partial\lambda}{\partial l}. \quad (76)$$

Remembering that $\delta\Phi + \lambda = \Phi_0 + \widetilde{\delta\phi}$ where Φ_0 is independent of l and that $dl = -L d\theta$, we obtain:

$$\delta E_\| = \frac{1}{L}\frac{\partial}{\partial\theta}(\widetilde{\delta\phi}) = \frac{1}{L}K_i e'(\theta), \quad (77)$$

where $e'(\theta)$ is the derivative of $e(\theta)$ with respect to θ. We have also performed the calculation for $n = 2$ and, in Figure 4, we plot the functions $e(\theta)$ and $e'(\theta)$ for these two values of n. As already mentioned, the results depend little upon n. Indeed, we note that whatever the value of n, $e'(\theta)$ goes to zero at the equator and has two symmetrical extrema off-equator. The parallel electric field, however, is more localized as n increases. In the following discussion, we will assume that we are located in the Northern hemisphere ($0 \le \theta \le \pi/2 \iff e'(\theta) < 0$). To go further and perform the energy integration, we

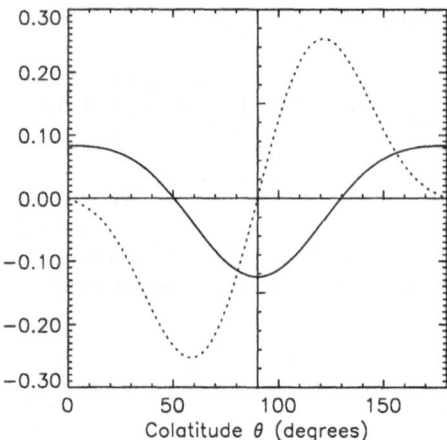

Fig. 4. Functions $e(\theta)$ plotted with solid line (resp. dashed line) for $n = 0$ (resp. $n=2$) and $e'(\theta)$ plotted with dotted line (resp. dotted-dashed line) for $n=0$ (resp. $n=2$).

need to simplify the term between brackets intervening in the expression (59) of K_i. We consider successively two extreme cases for which we are able to compute completely the parallel electric field. First, in Sect. 5.3, we assume that the curvature-gradient magnetic drift frequency is smaller than the frequency of the perturbation. Since $\overline{\omega}_d = k_y \overline{v_d}$, where $\overline{v_d}$ is proportional to the particle energy, this assumption is equivalent to consider that the particle energy is small or that the scale length of the perturbation in the y-direction $(2\pi/k_y)$ is large. Second, in Sect. 5.3 we assume that $\overline{\omega}_d$ is larger than ω which corresponds to high energy particles or to perturbations highly localized in longitude.

Low Magnetic Drift Limit ($\overline{\omega}_d < \omega$)

We assume that the particles, that ensure the quasi-neutrality, are low energy particles satisfying $\omega > \overline{\omega}_d$. This assumption means that the characteristic time for particles to escape from the perturbed region under the effect of their magnetic drift velocity $\overline{v}_d$, is long compared to the duration of the perturbation. For instance, at the geostationary orbit, thermal ions with 10 keV in the plasma sheet have a magnetic drift velocity of the order of 5 km/s. If the perturbed region has a width of 4 R_E in the y-direction, thermal ions need 90 min to cross the stretched region. This time exceeds the duration of the growth phase. Therefore, in this limit, the term between brackets of K_i

in (59) can be written:

$$\frac{\overline{\omega}_{di}\left(\overline{\omega}_{di} - \omega_{*i}\right)}{\omega\left(\omega + \overline{\omega}_{di}\right)} \simeq \frac{\overline{\omega}_{di}^2}{\omega^2}\left(1 - \frac{\omega_{*i}}{\overline{\omega}_{di}}\right). \tag{78}$$

From (65), the energy integration in (59) is performed and we obtain:

$$K_i = \frac{15}{4}\frac{T_e}{T_i}\overline{v}_{dthi}^2\left(1 - \frac{2}{5}\frac{u_{yi}}{\overline{v}_{dthi}}\right)\left(\frac{k_y^2\lambda_{0eq}}{\omega^2}\right). \tag{79}$$

In order to know whether the particles are accelerated towards the magnetic equator or towards the ionosphere, we compute an inverse Fourier transform in space and time for the parallel electric field (77). From the expressions (79) of K_i, (33) and (37), we obtain (see Appendix B):

$$\delta E_{\parallel}(L, y, \theta, t) = -e'(\theta)\left[\frac{15}{4}\frac{T_e}{T_i}\overline{v}_{dthi}^2\left(1 - \frac{2}{5}\frac{u_{yi}}{\overline{v}_{dthi}}\right)\right]\frac{\mu_0\delta j_m}{\sqrt{\pi}}\frac{L_c^2}{\Delta^2}\frac{\exp(\gamma t)}{\gamma}$$
$$\times\left(\zeta - \frac{1}{2}P\left(\tilde{W}(\zeta)\right) + \zeta^2 P\left(\tilde{W(\zeta)}\right)\right), \quad (80)$$

Close to midnight, $\zeta = y/\Delta < 1$, in this limit, the Fried-Conte function can be approximated by $P(\tilde{W}(\zeta)) \simeq -2\zeta + O(\zeta^3)$ and we obtain:

$$\delta E_{\parallel}(L, y, \theta, t) = -e'(\theta)\left[\frac{15}{2}\frac{T_e}{T_i}\overline{v}_{dthi}^2\left(1 - \frac{2}{5}\frac{u_{yi}}{\overline{v}_{dthi}}\right)\right]\frac{\mu_0\delta j_m}{\sqrt{\pi}}\frac{L_c^2}{\Delta^2}\frac{\exp(\gamma t)}{\gamma}$$
$$\times\left(\frac{y}{\Delta}\right)\left(1 - \left(\frac{y}{\Delta}\right)^2\right). \quad (81)$$

In the inverse limit $\zeta > 1$ (far away from the midnight meridian), the expansion of the Fried-Conte function is $-1/\zeta$ and the parallel electric field goes to zero:

$$\delta E_{\parallel}(L, y, \theta, t) = -e'(\theta)\left[\frac{15}{4}\frac{T_e}{T_i}\overline{v}_{dthi}^2\left(1 - \frac{2}{5}\frac{u_{yi}}{\overline{v}_{dthi}}\right)\right]$$
$$\times\frac{\mu_0\delta j_m}{\sqrt{\pi}}\frac{L_c^2}{\Delta^2}\frac{\exp(\gamma t)}{\gamma}\left(\frac{1}{2\zeta}\right). \quad (82)$$

From (81), we note that the parallel electric field changes sign between the dawn side and the dusk side and is maximum for $y = \pm\Delta/\sqrt{3}$, inside the stretched magnetic field line region $|y| < \Delta/\sqrt{2}$. In the following discussion, we will assume that we are located close to midnight ($\zeta < 1$). We distinguish between two cases:

Case A: $|u_y| < \frac{5}{2}|\overline{v}_d|$

$$\delta E_{\parallel} \simeq -\frac{15}{2}e'(\theta)\left[\frac{T_e}{T_i}\overline{v}_{dthi}^2\right]\frac{\mu_0\delta j_m}{\sqrt{\pi}}\frac{L_c^2}{\Delta^2}\frac{\exp(\gamma t)}{\gamma}\times\left(\frac{y}{\Delta}\right)\left(1 - \left(\frac{y}{\Delta}\right)^2\right). \tag{83}$$

The parallel electric field is negative (resp. positive), directed towards the equator (resp. towards the ionosphere) in the duskside (resp. in the dawn side). As a consequence, electrons will be accelerated towards the ionosphere in the dusk side and towards the equator at the dawn side.

Case B: $|u_y| > \frac{5}{2}|\bar{v}_d|$

$$\delta E_\parallel \simeq 3e'(\theta) \left[\frac{T_e}{T_i}\bar{v}_{dthi}u_y\right] \frac{\mu_0\delta j_m}{\sqrt{\pi}} \frac{L_c^2}{\Delta^2} \frac{\exp(\gamma t)}{\gamma} \times \left(\frac{y}{\Delta}\right)\left(1-\left(\frac{y}{\Delta}\right)^2\right). \quad (84)$$

In this case, the parallel electric field direction depends on the direction of the pressure gradient. In the magnetic tail where the pressure gradient is directed earthward, electrons are accelerated towards the equator in the dusk side and towards the ionosphere in the dawn side. Compared to case A, the situation is reversed.

On the other hand, in regions closer to the Earth, where the pressure gradient may be directed towards the tail, each of the above direction is reversed. Case B is presented in Figure 5 where we have taken into account the change of direction of the pressure gradient when we move closer to the Earth. The direction of the expected electron precipitations is summarized in Table (1).

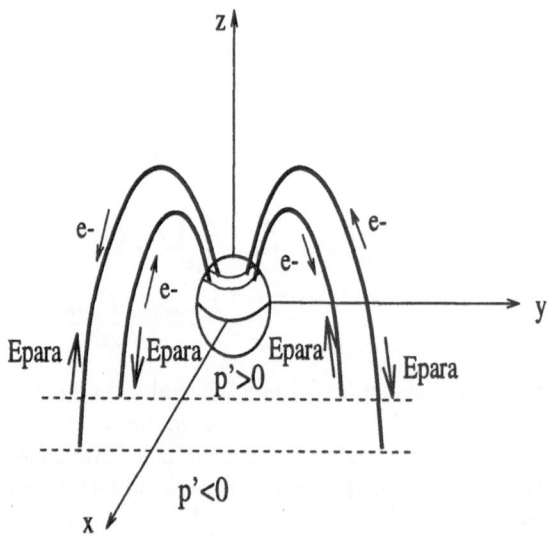

Fig. 5. Variation of the direction of the parallel electric field and of the electron precipitations as a function of y and of the pressure gradient (case B). The configuration is symmetrical about the magnetic equator.

	$\overline{\omega}_d < \omega$ et $y < 0$	
	$p' > 0$	$p' < 0$
$\overline{\omega}_d > \omega_*$	$\Longrightarrow$ Io	$\Longrightarrow$ Io
$\overline{\omega}_d < \omega_*$	$\Longrightarrow$ Io	$\Longrightarrow$ Eq

Table 1. Table of electron precipitations in the dusk side ($y < 0$) in the limit $\overline{\omega}_d < \omega$. ($\Longrightarrow$ Eq (resp. Io) means that electrons are accelerated equatorward (resp. towards the ionosphere)).

Large Magnetic Drift Limit ($\overline{\omega}_d > \omega$)

Now, we consider the opposite assumption $\overline{\omega}_d > \omega$, which corresponds to a very low frequency perturbation or to a very large k_y (an highly localized perturbation in longitude). In this limit, the term between brackets of K_i in (59) can be approximated by:

$$\frac{\overline{\omega}_{di}\left(\overline{\omega}_{di} - \omega_{*i}\right)}{\omega\left(\omega + \overline{\omega}_{di}\right)} \simeq \frac{\overline{\omega}_{di}}{\omega}\left(1 - \frac{\omega_{*i}}{\overline{\omega}_{di}}\right). \tag{85}$$

Therefore, after an integration on the energy E, the expression (59) of K_i, in the limit $\overline{\omega}_d > \omega$, writes:

$$K_i = \frac{3}{2}\frac{T_e}{T_i}\overline{v}_{dthi}\left(1 - \frac{2}{3}\frac{u_{yi}}{\overline{v}_{dthi}}\right)\left(\frac{k_y\lambda_{0eq}}{\omega}\right), \tag{86}$$

which gives for the parallel electric field, after an inverse Fourier transform in space and time (see Appendix B for the method):

$$\delta E_{\parallel}(L, y, \theta, t) = \frac{3}{2}e'(\theta)\frac{T_e}{T_i}\overline{v}_{dthi}\left(1 - \frac{2}{3}\frac{u_{yi}}{\overline{v}_{dthi}}\right)$$
$$\times \frac{\mu_0\delta j_m}{\sqrt{\pi}}\frac{L_c^2}{\Delta}\exp(\gamma t)\times\left(1 + \zeta P\left(\tilde{W}(\zeta)\right)\right). \tag{87}$$

Close to midnight, $\zeta = y/\Delta < 1$, we obtain:

$$\delta E_{\parallel}(L, y, \theta, t) = \frac{3}{2}e'(\theta)\frac{T_e}{T_i}\overline{v}_{dthi}\left(1 - \frac{2}{3}\frac{u_{yi}}{\overline{v}_{dthi}}\right)\frac{\mu_0\delta j_m}{\sqrt{\pi}}\frac{L_c^2}{\Delta}\exp(\gamma t)$$
$$\times\left(1 - 2\left(\frac{y}{\Delta}\right)^2\right). \tag{88}$$

In the opposite limit $\zeta > 1$ (far away from the midnight meridian), the expansion of the Fried-Conte function is $-1/\zeta$ and the parallel electric field goes to zero:

$$\delta E_{\parallel}(L, y, \theta, t) \simeq 0. \tag{89}$$

194   René Pellat et al.

Equation (88) shows that the parallel electric field does not change sign in the region where the magnetic field lines are stretched ($|y| < \Delta/\sqrt{2}$). In the following discussion, we will consider a location close to midnight ($\zeta < 1$). We distinguish again between two cases:

Case C: $|u_y| < \frac{3}{2}|\bar{v}_d|$

$$\delta E_{\parallel} \simeq \frac{3}{2} e'(\theta) \frac{T_e}{T_i} \bar{v}_{dthi} \frac{\mu_0 \delta j_m}{\sqrt{\pi}} \frac{L_c^2}{\Delta} \exp(\gamma t) \times \left(1 - 2\left(\frac{y}{\Delta}\right)^2\right). \tag{90}$$

The parallel electric field is positive around midnight, directed towards the ionosphere. Thus, electrons are accelerated towards the equator in the region where the magnetic field lines are more stretched. They are accelerated towards the ionosphere in the morning and evening sectors for $|y| > \Delta/\sqrt{2}$.

Case D: $|u_y| > \frac{3}{2}|\bar{v}_d|$

$$\delta E_{\parallel} \simeq -e'(\theta) \frac{T_e}{T_i} u_{yi} \frac{\mu_0 \delta j_m}{\sqrt{\pi}} \frac{L_c^2}{\Delta} \exp(\gamma t) \times \left(1 - 2\left(\frac{y}{\Delta}\right)^2\right). \tag{91}$$

In this last case, we find again a pressure gradient dependence of the parallel

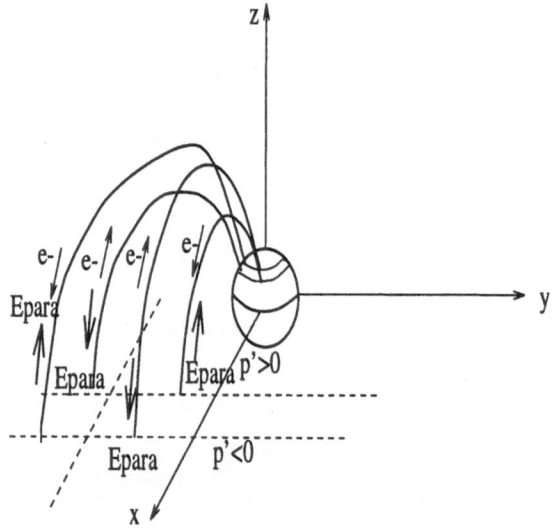

Fig. 6. Variations of the direction of the parallel electric field and of the electron precipitations in the dusk side as a function of the pressure gradient (case D). The dawn side configuration is obtained by symmetry around the noon/midnight meridian.

electric field. In the plasma sheet, where the pressure gradient is directed earthward (p' is negative), the parallel electric field is directed towards the equator and the electrons are accelerated towards the ionosphere around midnight. In the flanks, the situation is reversed. Closer to the Earth, the pressure may be directed tailward, then all the results are reversed; electrons move towards the ionosphere at midnight and towards the equator in the flanks (Figure 6). The directions of the expected electron motion are summarized in the Table (2).

	$\overline{\omega}_d > \omega$ et $|y| < \Delta/\sqrt{2}$	
	$p' > 0$	$p' < 0$
$\overline{\omega}_d > \omega_\star$	$\Longrightarrow$ Eq	$\Longrightarrow$ Eq
$\overline{\omega}_d < \omega_\star$	$\Longrightarrow$ Eq	$\Longrightarrow$ Io

Table 2. Table of electron precipitations in the region around the noon/midnight meridian ($|y| < \Delta/\sqrt{2}$) in the limit $\overline{\omega}_d > \omega$. ($\Longrightarrow$ Eq (resp. Io) means that electrons are accelerated equatorward (resp. toward the ionosphere)).

Discussion

The duration of the growth phase T_g is of order 30 minutes and the ion thermal energy is around 10 keV in the near Earth plasma sheet. Moreover, dual satellite studies suggest that the scale length of the perturbation associated with the formation of the current sheet during the growth phase can not be smaller than one Earth radius. Therefore, the assumption $\overline{\omega}_d \simeq \overline{v}_d/\Delta < \omega \simeq 2\pi/T_g$ seems reasonnable. Furthermore, from the azimuthal assymetry of the ion flux (between duskward/dawnward directions) measured aboard a geostationary spacecraft (e.g. [36]), we know that the scale length, L_P, of the ion pressure gradient is relatively small ($L_P \simeq 0.5 - 1$ R_E). Therefore, we have $u_y/\overline{v_d} \simeq L/L_p \simeq 6 - 10$ and we can consider that $|u_y| > |\overline{v_d}|$ close to the geosynchronous region. Thus, it seems that the most appropriate case for the substorm growth phase is case B, since in this case the particles that ensure the QNC verify $\overline{\omega}_d < \omega$ and $|u_y| > |\overline{v}_d|$. Therefore, we expect the parallel electric field to be directed towards the ionosphere in the duskside whereas it should be directed towards the magnetic equator in the dawn side. As a consequence, cold ionospheric electrons may be accelerated towards the equator at dusk, whereas hot plasma sheet electrons will move towards the ionosphere at dawn. This result is linked to the assumption that the perturbation associated with the magnetic field lines stretching is localized along the y-direction. The magnitude of the parallel electric field during the growth phase can be estimated. In (84), we assume (i) the time scale for the growth phase of about 30 minutes, (ii) the characteristic increase of the radial component of the magnetic field of 30 nT, (iii) the radial extent of the current sheet of 1 R_E, and (iv) the forcing current L_c is at 7 R_E.

Remind that the diamagnetic drift velocity is $u_y = p'/(qn_0) \simeq T/(qL_P B_{eq})$ and that the bounce averaged curvature-gradient magnetic drift velocity is $\overline{v}_{dth} = -2T/(qLB_{eq})$. We use the following parameters: 1 keV for the electron temperature and 10 keV for the ion temperature. The scale, Δ, of the perturbed region is taken to be 4 R_E, L_P, the scale length of the ion pressure gradient is 2 R_E, (much larger than values estimated in [36]), then the parallel electric field at the geostationnary orbit $L \simeq 6.6$ R_E , given by (84), is $E_{\parallel} \simeq 1.6 \times 10^{-5} \text{V.m}^{-1}$ with $e'(\theta) \simeq 0.1$. A smaller scale for the ion pressure gradient or/and a smaller Δ would result in a larger parallel electric field. For a distance between the ionosphere and the magnetic equator equal to $\pi L/2$, we obtain a field-aligned potential drop of a few hundred volts, which is far from negligible. We suggest that this parallel electric field produces a parallel drift velocity between electrons and ions. It has been shown elsewhere that the drifting electrons may generate electromagnetic waves, observed at substorm breakup [37] with frequencies of the order of the proton gyrofrequency ($F_{H+} \simeq 1$ Hz). More recently it has been shown in [38] that these waves are produced via a current driven instability when the parallel drift velocity between electrons and ions is large enough (a fraction of the ion thermal velocity). Since these waves are able to diffuse the electrons in a time corresponding to their bounce period (10 s), the parallel current will be disrupted which allows the magnetic reconfiguration observed in the near-Earth magnetotail (7-12 R_E) to occur.

    Present results seem to contradict those of Koskinen (compare Fig. 3 of [39] with our Fig. 3). But as Koskinen pointed out: "the parallel field of Fig. 3 is not the real electric field in the plasma. Due to their high parallel mobility charged particles quickly compensate the parallel component of the slowly evolving inductive field". In fact, during the growth phase, we have the following ordering $\omega_{be} > \omega_{bi} > \omega$. Thus, the field-aligned potential drop results from the differences between the bounce average responses of electrons and that of ions, owing to their different temperature. In the limit $T_e/T_i < 1$, this difference is small and therefore the field-aligned potential is small too. This is why the lowest order of the QNC gives $\delta\Phi + \lambda = constant$, which means that to the lowest order in (T_e/T_i), the inductive component of the parallel electric field $\partial A_{\parallel}/\partial t = \partial\lambda/\partial l$ is balanced by the electrostatic component $\partial\delta\Phi/\partial l$. Thus, the parallel electric field found by [39], from a non self-consistent approach, does not develop because the particle response tends to cancel it. Notice that the high parallel mobility, corresponding to large ion and electron temperatures, does not cancel the parallel electric field in a mirror geometry for low frequency perturbations. In the present paper, we have shown, for the simple case where $T_e < T_i$, that a parallel electric field proportional to T_e/T_i does develop. The present study shows the importance of ensuring QNC while imposing magnetic field variations. Finally, we point out we have solved the full Vlasov-Maxwell system of equations only in the quasi-static limit, notably the solution of Ampère's law is valid only in this limit. However, the results obtained from the lowest order in T_e/T_i of the QNC and to the first order are basically valid for low frequency perturbations $\omega < \omega_b$.

Although the above results allow a self-consistent description of the origin of parallel electric fields, there are some limitations that should be stressed. The present calculation has been carried out with the low β assumption since the 2D dipole has been used to solve the Ampère's law with a forcing current and to obtain the explicit expression for the parallel electric field. This limit is not *a priori* valid for the Earth's plasma sheet, especially for large radial distances. However, even with this restrictive assumption we have taken into account the pressure gradient effect to the lowest order. For higher values of β, we expect the general trend to be the same but another equilibrium should be used which would preclude any analytical approach.

Another limitation is the fact that we do not take into account the ionosphere/magnetosphere coupling. The QNC for the magnetotail plasma is likely to be modified by the presence of the ionospheric plasma. Keeping in mind the above limitations, the present model provides a self-consistent kinetic description of the generation of parallel electric fields in a mirror geometry for quasi-static perturbations ($\omega < k_\parallel v_A$ and $\omega < \omega_b$).

6 Conclusion

A detailed discussion has been given about the applicability of MHD to the description of active events occuring in the Earth's plasma sheet, before and at substorms, as well as in the solar corona, where Loops, Current Sheets, and Streamers develop. As long as collisions occur over times which are shorter than for other dynamical processes, MHD always applies. In a collisionless regime, however, we have shown that the applicability of ideal MHD is not granted and in particular is restricted to frequencies above the electron (and ion) bounce frequency. Thus against the conventional wisdom, in a mirror geometry, for a collisionless plasma, MHD is not a priori applicable to low frequency processes. We have presented a self-consistent kinetic approach of low frequency, quasi-static perturbations developing in a collisionless plasma confined by magnetic mirrors. Using the linear cyclotron and bounce averaged Vlasov equation obtained in [15], the QNC has been developed. To the lowest order in T_e/T_i ($T_e/T_i < 1$), the following results were described:

- The QNC imposes the existence of a component Φ_0, given by (13), of the perturbed electrostatic potential. This component Φ_0 is constant along the field line and varies in the azimuthal direction, thereby contributing to the azimuthal electric field.

- The presence of this electrostatic component does modify the plasma compression term in the MHD energy principle developed by Bernstein et al. [27]. Taking into account the existence of Φ_0, implied by the QNC, we have derived a kinetic variational form for the stochastic as well as for the adiabatic case. In the case of stochastic ion motion, we have shown that an MHD-like energy principle is recovered, thanks to the randomization of the pitch angles while ions cross the minimum B region (the equator) and we computed explicitly the plasma compression term.
- The potential electric field derived from QNC, has also direct consequences on the transport: it tends to reduce (partially shield) the effect of the inductive component of the electric field. In particular, this effect explains why no large bulk flows are found to be associated with large scale electromagnetic perturbations ($\tau > \tau_b$), such as substorm growth phase or long period oscillations (Pc5). Unlike what is done for the particle test and MHD approaches, the electrostatic component of the azimuthal electric field is not assumed; it is determined, in a self-consistent manner, by the response of the plasma to an externally applied perturbation. The existence of this electrostatic component Φ_0 is a purely kinetic effect. The total azimuthal electric field (14), which is the sum of these two components, varies in amplitude and direction, as a function of the position along the field line, which implies that the bounce-averaged transport of the particles (54) strongly depends on the pitch angle.
- The parallel electric field is null to the order $T_e/T_i < 1$ (see (11)), therefore the residual parallel electric field has been calculated from QNC developed to the order in T_e/T_i.

To the next order in T_e/T_i, an analytical calculation of the residual field-aligned potential drop has been presented. As long as we can assume that: $\overline{\omega}_d < \omega$ and $|\overline{v_d}| < |u_y|$, a field-aligned potential drop of a few hundred volts is found to develop as a consequence of the stretching of the magnetic field lines, during the substorm growth phase. The associated parallel electric field is directed towards the ionosphere in the dusk sector and towards the equator in the dawn sector. This parallel electric field is able to produce a parallel drift velocity between electrons and ions. In turn the drifting electrons generate electromagnetic waves, observed at magnetospheric substorm breakup [37] at frequencies of the order of the proton gyrofrequency ($F_{H+} \simeq 1$ Hz), via a current driven instability. We suggest that, thanks to their large amplitudes, these waves can:

- diffuse the electrons along the field line interrupting the bounce motion
- produce pitch angle and spatial diffusion of ions, thereby reducing/suppressing the pressure anisotropy and the pressure gradient responsible for the perpendicular current.

The diffusion of electrons being faster (10 s) than the diffusion of ions (100 s), the fast bounce motion of electrons is interrupted first. On the time scale of the electronic bounce period, the parallel current is disrupted and the system equilibrium is broken. Another consequence of the electron diffusion is that the electrostatic component Φ_0 is cancelled. The total perpendicular electric field is therefore no longer reduced and is equal to the large induced electric field component associated with the magnetic reconfiguration; the plasma moves due to the associated electric drift and the observed high convective flows can occur. Hence it seems that, in a collisionless plasma, a microscopic process (a current driven instability) controls the interruption of the current that was holding the current sheet. Notice that in this model there is no need for the formation of a neutral line; neutral line or points can develop as a consequence of the change of the system of electrical currents. Thus no particular role is played by a singular region ($B \simeq 0$), and there is no need for assuming an anomalous diffusion to initiate magnetic reconfiguration. The transformation of the magnetic energy stored into the CS into kinetic energy can occur everywhere in the CS, there is no need for a singular X-line topology.

A   Solution of the Linearized Ampère's law

Using the local field-aligned coordinates and with the gauge $\delta A_y = 0$, the curl of B gives

$$\nabla \times \delta B = \left\{ \frac{\partial}{\partial y} \left(-\frac{\partial \delta A_\psi}{\partial y} \right) - \frac{1}{JB} \frac{\partial}{\partial \chi} \left(\frac{1}{J} \left[\frac{\partial}{\partial \chi} \left(\frac{\delta A_\psi}{B} \right) \right. \right. \right.$$
$$\left. \left. - \frac{\partial}{\partial \psi} \left(JB\delta A_\| \right) \right] \right) \right\} e_\psi \tag{92}$$

$$+ \frac{1}{J} \left\{ \frac{\partial}{\partial \chi} \left(\frac{1}{B} \frac{\partial \delta A_\|}{\partial y} \right) - \frac{\partial}{\partial \psi} \left(JB \left(-\frac{\partial \delta A_\psi}{\partial y} \right) \right) \right\} e_y \tag{93}$$

$$+ \left\{ B \frac{\partial}{\partial \psi} \left(\frac{1}{J} \left[\frac{\partial}{\partial \chi} \left(\frac{\delta A_\psi}{B} \right) - \frac{\partial}{\partial \psi} \left(JB\delta A_\| \right) \right] \right) \right.$$
$$\left. - \frac{\partial}{\partial y} \left(-\frac{\partial \delta A_\|}{\partial y} \right) \right\} e_\|, \tag{94}$$

where J is the jacobian of the change of coordinates between the cartesian and the local frame (see also the appendix A of [20]). We assume that

$$\delta A(r, t) = \widehat{\delta A}(\psi, k_y, l, \omega) \exp\left(i(k_y y + \omega t) \right),$$

for the sake of simplicity we omit the $\widehat{}$ symbol and the exponential factor in the following formulas. Defining two variables $X = k_y \lambda / \omega$ and $\tilde{Y} = i k_y \delta A_\psi / B$ as [27], interchanging the partial derivatives, $\partial_\psi (JB\partial_l) = JB\partial_l \partial_\psi$, (see [20] for details about the interchange of the partial derivatives) when necessary, the Ampère's law writes:

$$-ik_y \tilde{Y}B - \frac{1}{ik_y}\frac{\partial}{\partial l}\left(B\frac{\partial \tilde{Y}}{\partial l} - B\frac{\partial^2 X}{\partial l \partial \psi}\right) = 0, \tag{95}$$

$$B\frac{\partial}{\partial l}\left(\frac{1}{B}\frac{\partial X}{\partial l}\right) + \frac{1}{J}\frac{\partial}{\partial \psi}\left(JB^2\tilde{Y}\right) = \mu_0 \delta j_{yext}, \tag{96}$$

$$\frac{B}{ik_y}\frac{\partial}{\partial \psi}\left(B\frac{\partial \tilde{Y}}{\partial l} - B\frac{\partial^2 X}{\partial l \partial \psi}\right) - ik_y \frac{\partial X}{\partial l} = 0. \tag{97}$$

One can show that (97) can be obtained from (95) and (96), therefore the system is reduced to these two latter equations. Inserting (95) to (96), we obtain

$$B\frac{\partial}{\partial l}\left(\frac{1}{B}\frac{\partial X}{\partial l}\right) + \frac{1}{Jk_y^2}\frac{\partial}{\partial \psi}\left(JB\frac{\partial}{\partial l}\left(B\frac{\partial \tilde{Y}}{\partial l} - B\frac{\partial^2 X}{\partial l \partial \psi}\right)\right) = \mu_0 \delta j_{yext}. \tag{98}$$

Again, we interchange the partial derivatives $(\partial_\psi(JB\partial_l) = JB\partial_l\partial_\psi)$, divide by B and integrate along the field line which yields

$$\left(\frac{1}{B}\frac{\partial X}{\partial l}\right) + \frac{1}{k_y^2}\frac{\partial}{\partial \psi}\left(B\frac{\partial \tilde{Y}}{\partial l} - B\frac{\partial^2 X}{\partial l \partial \psi}\right) = \mu_0 \int \frac{dl}{B}\delta j_{yext}. \tag{99}$$

Therefore we have to solve the following system of equations:

$$\tilde{Y} = \frac{1}{k_y^2 B}\frac{\partial}{\partial l}\left(B\frac{\partial \tilde{Y}}{\partial l} - B\frac{\partial^2 X}{\partial l \partial \psi}\right), \tag{100}$$

$$\left(\frac{1}{B}\frac{\partial X}{\partial l}\right) + \frac{1}{k_y^2}\frac{\partial}{\partial \psi}\left(B\frac{\partial \tilde{Y}}{\partial l} - B\frac{\partial^2 X}{\partial l \partial \psi}\right) = \mu_0 \int \frac{dl}{B}\delta j_{yext}. \tag{101}$$

To go further, we assume *a priori* that $\partial \tilde{Y}/\partial l < \partial^2 X/\partial l \partial \psi$ (and will check it afterwards) which corresponds to assume that the variation of $\delta B_\parallel$ along the field line is smaller than the variation of δB_ψ in the radial direction. This assumption is well adapted to the choice of an external current highly localized in the radial direction. The system becomes:

$$\tilde{Y} = -\frac{1}{k_y^2 B}\frac{\partial}{\partial l}\left(B\frac{\partial^2 X}{\partial l \partial \psi}\right), \tag{102}$$

$$\left(\frac{1}{B}\frac{\partial X}{\partial l}\right) - \frac{1}{k_y^2}\frac{\partial}{\partial \psi}\left(B\frac{\partial^2 X}{\partial l \partial \psi}\right) = \mu_0 \int \frac{dl}{B}\delta j_{yext}. \tag{103}$$

Defining a new variable, $U = \partial X/\partial l$, (103) becomes:

$$\frac{\partial^2 U}{\partial \psi^2} + \frac{\partial \ln JB^2}{\partial \psi} \frac{\partial U}{\partial \psi}$$

$$- \frac{k_y^2}{B^2} U \left(1 - \frac{B}{k_y^2} \frac{\partial}{\partial \psi} \left(\frac{1}{J} \frac{\partial JB}{\partial \psi}\right)\right) = -\frac{\mu_0 k_y^2}{B} \int \frac{dl}{B} \delta j_{yext}. \quad (104)$$

Assuming that the perturbation varies faster in the ψ direction than the equilibrium, the second term of (104) can be neglected. Moreover, in the limit $|k_y|L > 1$ the term between parenthesis is equivalent to unity. Therefore, (104) can be rewritten in a classical form of a linear second order differential equation:

$$\frac{\partial^2 U}{\partial \psi^2} - \frac{k_y^2}{B^2} U = S(\psi, k_y, l, \omega), \quad (105)$$

where $S(\psi, k_y, l, \omega) = -(\mu_0 k_y^2/B) \int (dl/B \delta j_{ext}(\psi, k_y, l, \omega))$ is the forcing term. To solve (105), we have to build a Green function from the solutions of the homogeneous equation; $U_1 = \exp\left(|k_y| \int d\psi/B\right)$ and $U_2 = \exp\left(-|k_y| \int d\psi/B\right)$ (see e.g. [40]). Taking into account the properties of the 2D dipole model it is more convenient to use the variables L and θ defined by (26). While L and θ are not strictly speaking independent variables, we can consider them as such in the limit where we neglect the variations of the equilibrium compared to those of the perturbation, these variables allow us to easily solve (105), which becomes:

$$\frac{\partial^2 U}{\partial L^2} - (k_y^2 \sin^4 \theta)U = S(L, k_y, \theta, \omega), \quad (106)$$

where $S(L, k_y, \theta, \omega) = \mu_0 k_y^2 L \sin^2 \theta \int d\theta \sin^2 \theta \delta j_{ext}(L, k_y, \theta, \omega)$. Now, the solutions of the homogeneous equation become $U_1 = \exp|k_y|L \sin^2 \theta$ and $U_2 = \exp -|k_y|L \sin^2 \theta$. The Green function is defined by:

$$G(L, L_0) = \begin{cases} \frac{U_1(L)U_2(L_0)}{W(L_0)} = \frac{\exp\left(|k_y|(L-L_0)\sin^2 \theta\right)}{W(L_0)}, & \text{for } 0 \leq L \leq L_0; \\ \\ \frac{U_1(L_0)U_2(L)}{W(L_0)} = \frac{\exp\left(-|k_y|(L-L_0)\sin^2 \theta\right)}{W(L_0)}, & \text{for } L_0 \leq L < \infty, \end{cases}$$

where $W(L_0) = -2|k_y| \sin^2 \theta$, is the Wronskian of U_1 and U_2. The Green function becomes:

$$G(L, L_0) = -\frac{\exp\left(-|k_y(L-L_0)|\sin^2 \theta\right)}{2|k_y| \sin^2 \theta}. \quad (107)$$

The complete solution of (106) writes (omitting to specify all the dependences)

$$U(L) = \int_0^\infty dL_0 G(L, L_0) S(L_0), \quad (108)$$

which yields (specifying all the spatio-temporal dependences)

$$
U(L, \theta, k_y, \omega) = -\frac{\mu_0 |k_y|}{2} \int_0^\infty dL_0 L_0 \exp\left(-|k_y(L - L_0)| \sin^2 \theta\right)
$$
$$
\times \int d\theta \sin^2 \theta \delta j_{yext}(L_0, \theta, k_y, \omega). \quad (109)
$$

Taking into account the expression (30) of the external current, the complete solution for this class of current is

$$
U_n(L, k_y, \theta, \omega) = -\frac{\mu_0 |k_y| \delta j_{eq}(k_y, \omega)}{2} L_c^2 \exp\left(-|k_y(L - L_c)| \sin^2 \theta\right)
$$
$$
\times \left(\sin^{2n+1} \theta \cos \theta + C(k_y, \psi)\right), \quad (110)
$$

where we have used the following integral:

$$
\int d\theta \sin^{2n+2} \theta \left((2n + 1) \cot^2 \theta - 1\right) = \sin^{2n+1} \theta \cos \theta + C(k_y, \psi)
$$

We impose $U = \partial X / \partial l = \delta B_\psi = 0$ at the equator ($\theta = \pi/2$) which implies $C = 0$. From (110), we obtain

$$
X_n(L, k_y, \theta, \omega) = \frac{\mu_0 |k_y| \delta j_{eq}(k_y, \omega)}{4} L_c^2 L \int du u^n \exp(Zu), \quad (111)
$$

where we have defined $u = \sin^2 \theta$ and $Z = -|k_y(L - L_c)|$. The solution becomes:

$$
X_n(L, k_y, \theta, \omega) = \frac{\mu_0 |k_y| \delta j_{eq}(k_y, \omega)}{4} L_c^2 L \left(\frac{u^n}{Z}\right.
$$
$$
\left. + \sum_{k=1}^n (-1)^k n(n - 1)...(n - k + 1)\frac{u^{n-k}}{Z^{k+1}}\right) \exp(Zu), \quad (112)
$$

where we have used the integral $\int du u^n \exp(Zu) = [u^n/Z + \sum_{k=1}^n (-1)^k n(n - 1)...(n - k + 1)u^{n-k}/Z^{k+1}] \exp(Zu)$. Now, we can solve the second equation (102) of the system to obtain $\tilde{Y}$. With the variables L and θ, and in the limit $|k_y|L > 1$, (102) becomes:

$$
\tilde{Y} = \frac{1}{k_y^2 BL} \frac{\partial}{\partial \theta} \left(B \frac{L^2}{\hat{D}} \frac{\partial U}{\partial L}\right) \quad (113)
$$

After some algebra, we obtain

$$
\tilde{Y} = \frac{\mu_0 \delta j_{eq}(k_y, \omega)}{2} L_c^2 \frac{L}{\hat{D}} u^{n+1} \left((2n + 1) - (2n + 2)u + 2Zu(1 - u)\right)
$$
$$
\times [H(L - L_c) - H(-(L - L_c))] \exp(Zu), \quad (114)
$$

where H stands for the Heaviside function. As claimed above, we verify *a posteriori* that the assumption $\partial \tilde{Y}/\partial l < \partial^2 X/\partial l \partial \psi$ is valid. Indeed, we have

$$\frac{\partial \tilde{Y}/\partial l}{\partial^2 X/\partial l \partial \psi} \simeq \frac{\partial \tilde{Y}/\partial l}{\partial U/\partial \psi} \simeq \frac{1}{k_y^2 L^2}. \tag{115}$$

Thus, the above assumption is valid in the limit $|k_y|L > 1$. Then, we take the limit $|Z| = |k_y(L - L_c)| < 1$, which implies that the wave length of the perturbation is larger than the distance between the current and the location where the solution is calculated. In this limit, one can show that X writes:

$$X_n(L, k_y, \theta, \omega) \simeq \frac{\mu_0 |k_y| \delta j_{eq}(k_y, \omega)}{4} L_c^2 L \left(\frac{n!}{Z^{n+1}} + \frac{u^{n+1}}{n+1} + O(Z) \right). \tag{116}$$

Finally, we obtain for λ:

$$\lambda_n(L, k_y, \theta, \omega) \simeq \lambda_{neq} \left(c_n(L, k_y) + (\sin^2 \theta)^{n+1} + O(-|k_y(L - L_c)|) \right), \tag{117}$$

where we have defined $c_n(L, k_y) = (n + 1)!/(-|k_y(L - L_c)|)^{n+1}$ and $\lambda_{neq} = 1/(4(n + 1))\mu_0(|k_y|/k_y)\omega \delta j_{eq}(k_y, \omega)L_c^2 L$. The second variable, $\tilde{Y}$, becomes

$$\tilde{Y} = 2(n + 1)\lambda_{neq} \frac{k_y}{|k_y|\omega \hat{D}} \sin^{2n+2} \theta \left((2n + 1) - (2n + 2)\sin^2 \theta\right)$$

$$\times \left[H(L - L_c) - H\left(-(L - L_c)\right)\right]. \tag{118}$$

Therefore, in the limit $|k_y|L > 1$ and $|k_y(L - L_c)| < 1$, the perturbed components of the magnetic field write:

$$\delta B_\psi = U_n(L, \theta) = -\frac{\mu_0 |k_y| \delta j_{eq}(k_y, \omega)}{2} L_c^2 \left(\sin^{2n+1} \theta \cos \theta\right) \tag{119}$$

$$\delta B_\parallel = -B\tilde{Y} = -\frac{\mu_0 \delta j_{eq}(k_y, \omega)}{2} \frac{L_c^2}{L} \sin^{2n} \theta \left((2n + 1) - (2n + 2)\sin^2 \theta\right)$$

$$\times \left[H(L - L_c) - H\left(-(L - L_c)\right)\right] \tag{120}$$

B   Inversion of the Fourier Transform

In this appendix, we do not omit the Fourier notations therefore, we have

$$\hat{U} = \frac{\partial \hat{X}}{\partial l} = \widehat{\delta B_\psi}, \tag{121}$$

and the real component δB_ψ writes

$$\delta B_\psi(L, y, \theta, t) = \frac{1}{2\pi} \int_{-\infty}^{+\infty} d\omega \int_{-\infty}^{+\infty} dk_y \widehat{\delta B_\psi}(L, k_y, \theta, \omega)$$
$$\times \exp\left(i\left(k_y y + \omega t\right)\right), \tag{122}$$

$$\delta B_\psi(L, y, \theta, t) = -\frac{\mu_0}{2} f(\theta) \frac{1}{2\pi} \int_{-\infty}^{+\infty} d\omega \int_{-\infty}^{+\infty} dk_y |k_y| \widehat{\delta j_{eq}}(k_y, \omega)$$
$$\times \exp\left(i\left(k_y y + \omega t\right)\right), \tag{123}$$

where $f(\theta) = L_c^2 \sin^{2n+1}\theta \cos\theta$. The ω integration is straightforward and gives

$$\delta B_\psi(L, y, \theta, t) = -\frac{\mu_0}{2\sqrt{2\pi}} f(\theta) \int_{-\infty}^{+\infty} dk_y |k_y| \widehat{\delta j_{eq}}(k_y, t) \exp\left(ik_y y\right). \tag{124}$$

For the k_y integration, we note that :

$$|k_y| = k_y \, sign(k_y) \tag{125}$$

$$= \frac{ik_y}{\pi} FT\left(P\left(\frac{1}{y}\right)\right) \tag{126}$$

$$= \frac{1}{\pi} FT\left\{\frac{\partial}{\partial y}\left[P\left(\frac{1}{y}\right)\right]\right\}. \tag{127}$$

where FT denotes a one dimension Fourier transform and P the Cauchy principal value. The perturbed magnetic field component becomes

$$\delta B_\psi = -\frac{\mu_0}{2} f(\theta) \int_{-\infty}^{+\infty} dk_y \left(\frac{1}{\pi} FT\left\{\frac{\partial}{\partial y}\left[P\left(\frac{1}{y}\right)\right]\right\}\right)$$
$$\times FT\left(\delta j_{eq}(y, t)\right) \exp\left(ik_y y\right). \tag{128}$$

Moreover, the convolution theorem gives :

$$\widehat{f * g} = \widehat{f} \cdot \widehat{g} \Longleftrightarrow f * g = FT^{-1}\left(\widehat{f} \cdot \widehat{g}\right). \tag{129}$$

Therefore, we obtain

$$\delta B_\psi = -\frac{\mu_0}{2\pi} f(\theta) \int_{-\infty}^{+\infty} dy' \delta j_{eq}(y', t) P\left(-\frac{1}{(y - y')^2}\right). \tag{130}$$

Finally, we can write

$$\delta B_\psi = \frac{\mu_0}{2\pi} f(\theta) P\left(\int_{-\infty}^{+\infty} dy' \frac{\delta j_{eq}(y', t)}{(y - y')^2}\right). \tag{131}$$

Another example, for the parallel electric field, we have

$$\widehat{E}_{\|}(L, k_y, \theta, \omega) = \frac{1}{L} e'(\theta) \left[\frac{15}{4} \frac{T_e}{T_i} \overline{v}_{dthi}^2 \left(1 - \frac{2}{5} \frac{u_{yi}}{\overline{v}_{dthi}} \right) \right] \frac{k_y^2}{\omega^2} \widehat{\lambda}_{neq}(k_y, \omega),$$

$$(132)$$

and the real component $\delta E_{\|}$ writes

$$\delta E_{\|}(L, y, \theta, t) = \frac{1}{2\pi} \int_{-\infty}^{+\infty} d\omega \int_{-\infty}^{+\infty} dk_y \delta \widehat{E}_{\|}(L, k_y, \theta, \omega) \exp\left(i\left(k_y y + \omega t \right) \right)$$

$$(133)$$

$$= \frac{1}{L} e'(\theta) \left[\frac{15}{4} \frac{T_e}{T_i} \overline{v}_{dthi}^2 \left(1 - \frac{2}{5} \frac{u_{yi}}{\overline{v}_{dthi}} \right) \right] \qquad (134)$$

$$\times \frac{1}{2\pi} \int_{-\infty}^{+\infty} d\omega \int_{-\infty}^{+\infty} dk_y \frac{k_y^2}{\omega^2} \widehat{\lambda}_{neq}(k_y, \omega) \exp\left(i\left(k_y y + \omega t \right) \right).$$

$$(135)$$

Using the expression of $\lambda_{neq} = 1/(4(n+1))\mu_0(|k_y|/k_y)\omega \delta j_{eq}(k_y, \omega)L_c^2 L$ for $n = 0$, we write

$$\delta E_{\|}(L, y, \theta, t) = e'(\theta) \left[\frac{15}{4} \frac{T_e}{T_i} \overline{v}_{dthi}^2 \left(1 - \frac{2}{5} \frac{u_{yi}}{\overline{v}_{dthi}} \right) \right] \frac{L_c^2 \mu_0}{4}$$

$$\times \frac{1}{2\pi} \int_{-\infty}^{+\infty} d\omega \int_{-\infty}^{+\infty} dk_y \frac{k_y}{\omega} |k_y| \widehat{\delta j}_{eq}(k_y, \omega) \exp\left(i\left(k_y y + \omega t \right) \right). \quad (136)$$

The ω integration is straightforward and gives

$$\delta E_{\|}(L, y, \theta, t) = e'(\theta) \left[\frac{15}{4} \frac{T_e}{T_i} \overline{v}_{dthi}^2 \left(1 - \frac{2}{5} \frac{u_{yi}}{\overline{v}_{dthi}} \right) \right] \frac{L_c^2 \mu_0}{4}$$

$$\times \frac{1}{\sqrt{2\pi}} \int_{-\infty}^{+\infty} dk_y ik_y |k_y| \int dt \widehat{\delta j}_{eq}(k_y, t) \exp\left(ik_y y \right). \quad (137)$$

For the k_y integration, we note that :

$$ik_y |k_y| = k_y^2 i \, \text{sign}(k_y), \qquad (138)$$

$$= -\frac{k_y^2}{\pi} FT \left(P\left(\frac{1}{y} \right) \right), \qquad (139)$$

$$= \frac{1}{\pi} FT \left\{ \frac{\partial^2}{\partial y^2} \left[P\left(\frac{1}{y} \right) \right] \right\}. \qquad (140)$$

The perturbed magnetic field component becomes

$$\delta E_{\parallel}(L,y,\theta,t) = e'(\theta) \left[\frac{15}{4} \frac{T_e}{T_i} \overline{v}_{dthi}^2 \left(1 - \frac{2}{5} \frac{u_{yi}}{\overline{v}_{dthi}} \right) \right] \frac{L_c^2 \mu_0}{4}$$

$$\times \int_{-\infty}^{+\infty} dk_y \left(\frac{1}{\pi} FT \left\{ \frac{\partial^2}{\partial y^2} \left[P\left(\frac{1}{y} \right) \right] \right\} \right) \cdot FT \left(\int dt \delta j_{eq}(y,t) \right) \exp\left(ik_y y \right). \tag{141}$$

Moreover, the convolution theorem gives :

$$\widehat{f * g} = \hat{f} \cdot \hat{g} \Longleftrightarrow f * g = FT^{-1} \left(\hat{f} \cdot \hat{g} \right) \tag{142}$$

Therefore, we obtain

$$\delta E_{\parallel}(L,y,\theta,t) = e'(\theta) \left[\frac{15}{4} \frac{T_e}{T_i} \overline{v}_{dthi}^2 \left(1 - \frac{2}{5} \frac{u_{yi}}{\overline{v}_{dthi}} \right) \right] \frac{L_c^2 \mu_0}{4\pi}$$

$$\times \int_{-\infty}^{+\infty} dy' \left(\int dt \delta j_{eq}(y',t) \right) P\left(\frac{2}{(y-y')^3} \right). \tag{143}$$

Finally, we can write

$$\delta E_{\parallel}(L,y,\theta,t) = e'(\theta) \left[\frac{15}{4} \frac{T_e}{T_i} \overline{v}_{dthi}^2 \left(1 - \frac{2}{5} \frac{u_{yi}}{\overline{v}_{dthi}} \right) \right] \frac{L_c^2 \mu_0}{2\pi}$$

$$\times P\left(\int_{-\infty}^{+\infty} dy' \frac{\int dt \delta j_{eq}(y',t)}{(y-y')^3} \right). \tag{144}$$

Using the expression of $\delta j_{eq}(y,t)$:

$$\delta j_{eq}(y,t) = \delta j_m \exp\left(-y^2/\Delta^2 \right) \exp \gamma t, \tag{145}$$

with the same previous variables $Y = y'/\Delta$ and $\zeta = y/\Delta$, the parallel electric field becomes:

$$\delta E_{\parallel}(L,y,\theta,t) = e'(\theta) \left[\frac{15}{4} \frac{T_e}{T_i} \overline{v}_{dthi}^2 \left(1 - \frac{2}{5} \frac{u_{yi}}{\overline{v}_{dthi}} \right) \right] \frac{L_c^2 \mu_0 \delta j_m}{2\pi \gamma \Delta^2} \exp(\gamma t)$$

$$\times P\left(\int_{-\infty}^{+\infty} dY \frac{\exp\left(-Y^2/\Delta^2 \right)}{(\zeta - Y)^3} \right), \tag{146}$$

where we have performed the temporal integration. Then, after two integrations by parts, we obtain:

$$\delta E_{\parallel}(L,y,\theta,t) = -e'(\theta) \left[\frac{15}{4} \frac{T_e}{T_i} \overline{v}_{dthi}^2 \left(1 - \frac{2}{5} \frac{u_{yi}}{\overline{v}_{dthi}} \right) \right] \frac{\mu_0 \delta j_m}{\sqrt{\pi}} \frac{L_c^2}{\Delta^2} \frac{\exp(\gamma t)}{\gamma}$$

$$\times P\left(\zeta - \frac{\tilde{W}(\zeta)}{2} + \zeta^2 \tilde{W}(\zeta) \right). \tag{147}$$

C   Nonlocal Terms

Here, we compute the first non-local term $\overline{\omega_d \lambda}/\omega$ of δf. In the field aligned coordinates, the expression of the magnetic curvature-gradient drift is [20]:

$$v_d = -\frac{mv_\parallel}{qJB}\frac{\partial}{\partial \psi}\left(JBv_\parallel\right), \tag{148}$$

After the ψ derivative, we obtain:

$$v_d = -\frac{1}{q}\frac{\partial}{\partial \psi}\left(\frac{1}{2}mv_\parallel^2\right) - \frac{mv_\parallel^2}{q}\frac{\partial}{\partial \psi}\left(\ln JB\right),$$

replacing $1/2mv_\parallel^2$ by $E - \mu B$ and taking into account the conservation of the kinetic energy E (static magnetic field and absence of electrostatic field) and conservation of the magnetic moment μ, in the adiabatic regime, we obtain:

$$v_d = \frac{1}{q}\mu\frac{\partial B}{\partial \psi} + \frac{mv_\parallel^2}{qB}\frac{\partial B}{\partial \psi},$$

where $J = 1/B^2$. Indeed, from the definition of the Jacobian $J = (1/B^2) \times \exp\left(-\int d\psi(\nabla \times B)\cdot e_y/B^2\right)$ [27], we have:

$$\frac{\partial \ln J}{\partial \psi} = -2\frac{\partial \ln B}{\partial \psi} - \frac{\mu_0 p'}{B^2},$$

which can be approximated by

$$\simeq -\frac{2}{LB} - \frac{2\mu_0 p}{L_p B^3},$$

where L_p is the scale length of the pressure gradient and L is the scale length of the magnetic field gradient. Noticing that $\beta = 2\mu_0 p/B^2$, the above relation becomes:

$$\frac{\partial \ln J}{\partial \psi} \simeq -\frac{2}{LB}\left(1 + \frac{1}{4}\beta\frac{L}{L_p}\right).$$

In the limit $\beta \ll 1$ even if the scale length of the pressure gradient is small $(L/L_p \gg 1)$, we can still get $\beta L/L_p \ll 1$ and write:

$$\frac{\partial \ln J}{\partial \psi} \simeq -\frac{2}{B}\frac{\partial B}{\partial \psi}, \tag{149}$$

which gives $J = 1/B^2$.

Now, we have to compute the ψ derivative of B, keeping χ constant. From (27), we obtain:

$$\frac{\partial B}{\partial \psi} = -\frac{2}{L}. \tag{150}$$

The purely magnetic drift becomes:

$$v_d = -\frac{2}{qL}\left[\mu + \frac{mv_\parallel^2}{B}\right]. \tag{151}$$

The bounce average of the product $v_d\lambda$ gives:

$$\overline{\omega_d\lambda} = \frac{1}{\tau_b}\oint \frac{dl}{|v_\parallel|}\left[-\frac{2k_y}{qL}\left(\mu + \frac{mv_\parallel^2}{B}\right)\right]\lambda. \tag{152}$$

With the definition of λ given in (33), we obtain

$$\overline{\omega_d\lambda} = \overline{\omega}_d c_n \lambda_{neq} - \frac{1}{\tau_b}\oint \frac{dl}{|v_\parallel|}\left[-\frac{2k_y}{qL}\left(\mu + \frac{mv_\parallel^2}{B}\right)(\sin^2\theta)^{n+1}\right]\lambda_{neq}. \tag{153}$$

In (153), T_1, the first term of the expression between bracket becomes:

$$T_1 = -\frac{2k_y}{qL}\mu\lambda_{neq}\frac{1}{\tau_b}\oint \frac{dl}{|v_\parallel|}(\sin^2\theta)^{n+1}, \tag{154}$$

The parallel velocity is given by:

$$|v_\parallel| = \sqrt{\frac{2}{m}(E - \mu B)}, \tag{155}$$

and from (25), we obtain:

$$|v_\parallel| = \sqrt{\frac{2E}{m}}\sqrt{\left(1 - \frac{\mu B_{eq}}{E\sin^2\theta}\right)}. \tag{156}$$

Then, the term T_1 becomes:

$$T_1 = \frac{8k_y}{qv}\mu\lambda_{neq}\frac{1}{\tau_b}\int_{\frac{\pi}{2}}^{\theta_m}\frac{d\theta(\sin^2\theta)^{n+1}}{\left(1 - \frac{\mu B_{eq}}{E\sin^2\theta}\right)^{1/2}}, \tag{157}$$

where v is the velocity, θ_m is the colatitude of the mirror point and we have used that $l = L(\pi/2 - \theta)$ so $dl = -Ld\theta$. Taking $Z = \cos\theta/A$ with $A = 1 - \mu B_{eq}/E$, we obtain:

$$T_1 = -\frac{8k_y}{qv}\mu\lambda_{neq}\frac{4}{\tau_r}\int_0^1 \frac{dZ}{(1 - Z^2)^{1/2}}(1 - AZ^2)^{n+1}. \tag{158}$$

Using the binomial formula, we obtain:

$$T_1 = -\frac{8k_y}{qv}\mu\lambda_{neq}\frac{4}{\tau_r}\int_0^1 \frac{dZ}{(1-Z^2)^{1/2}}\sum_{k=0}^{n+1}(-1)^k C_{n+1}^k A^k Z^{2k}, \tag{159}$$

$$= -\frac{8k_y}{qv}\mu\lambda_{neq}\sum_{k=0}^{n+1}(-1)^k C_{n+1}^k A^k \frac{4}{\tau_r}\int_0^1 \frac{dZ}{(1-Z^2)^{1/2}}Z^{2k}. \tag{160}$$

We give the following relation:

$$\int_0^1 \frac{dZ\, Z^{2k}}{(1-Z^2)^{1/2}} = \frac{\pi}{2}\frac{(2k-1)!!}{(2k)!!}\ ,\ \text{for } k \geq 0, \tag{161}$$

and perform the Z integration:

$$T_1 = -\frac{2k_y}{qL}\mu\lambda_{neq}\sum_{k=0}^{n+1}(-1)^k C_{n+1}^k A^k \frac{(2k-1)!!}{(2k)!!}. \tag{162}$$

Following the same approach, the second term T_2 of (153) gives:

$$T_2 = \frac{1}{\tau_b}\oint \frac{dl}{|v_\parallel|}\left[-\frac{2k_y}{qL}\left(\frac{mv_\parallel^2}{B}\right)\right](\sin^2\theta)^{n+1}, \tag{163}$$

$$= -\frac{2k_y E}{qLB_{eq}}\lambda_{neq}A\sum_{k=0}^{n+1}(-1)^k C_{n+1}^k A^k \frac{(2k-1)!!}{(2k)!!}\left(\frac{1}{k+1}\right). \tag{164}$$

Summing all terms, we obtain:

$$\overline{\omega_d\lambda} = \overline{\omega}_d\lambda_{neq}\left[c_n\right.$$
$$\left. + 2\sum_{k=0}^{n+1}(-1)^k C_{n+1}^k \frac{(2k-1)!!}{(2k+2)!!}\left(1-\frac{\mu B_{eq}}{E}\right)^k\left(1+k\frac{\mu B_{eq}}{E}\right)\right], \tag{165}$$

where $\overline{\omega}_d = -2k_y E/(qLB_{eq})$.

D Calculation of the Constant Part Φ_0 of the Perturbed Electrostatic Potential

Thanks to the simplicity of the 2D dipole, we are able to compute completely Φ_0. Remembering that, in the limit $|k_y|L \gg 1$, Φ_0 writes:

$$\Phi_0 = -\frac{\int \frac{dl}{B}\left[(\frac{q_i}{m_i})^2\int\frac{4\pi BdEd\mu}{|v_\parallel|}f_{0i}\times\left[\frac{\omega_{*i}-\overline{\omega}_{di}}{\omega+\overline{\omega}_{di}}\left(\frac{\overline{\omega}_{di}\lambda}{\omega}\right)\right]\right]}{\int \frac{dl}{B}\left[(\frac{q_i}{m_i})^2\int\frac{4\pi BdEd\mu}{|v_\parallel|}f_{0i}\times\left[\frac{\overline{\omega}_{di}(\overline{\omega}_{di}-\omega_{*i})}{\omega(\omega+\overline{\omega}_{di})}\right]\right]}. \tag{166}$$

Using (165), the numerator noted N of (166) becomes:

$$N = \int \frac{dl}{B}\left[(\frac{q_i}{m_i})^2 \int \frac{4\pi B dE d\mu}{|v_\|} f_{0i}\left[\frac{\overline{\omega}_{di}(\omega_{*i} - \overline{\omega}_{di})}{\omega(\omega + \overline{\omega}_{di})}\lambda_{neq}\left(c_n\right.\right.\right.$$

$$\left.\left.\left. + 2\sum_{k=0}^{n+1}(-1)^k C_{n+1}^k \frac{(2k-1)!!}{(2k+2)!!}\left(1 - \frac{\mu B_{eq}}{E}\right)^k \left(1 + k\frac{\mu B_{eq}}{E}\right)\right)\right]\right].$$

Using again the binomial formula, we can write:

$$N = \int \frac{dl}{B}\left[(\frac{q_i}{m_i})^2 \int 4\pi dE f_{0i} \times \left[\frac{\overline{\omega}_{di}(\omega_{*i} - \overline{\omega}_{di})}{\omega(\omega + \overline{\omega}_{di})}\lambda_{neq}\left(c_n\right.\right.\right.$$

$$\left.\left.\left. + 2\sum_{k=0}^{n+1}(-1)^k C_{n+1}^k \frac{(2k-1)!!}{(2k+2)!!}\sum_{j=0}^{k}(-1)^j C_k^j \left(\frac{\mu B_{eq}}{E}\right)^j \left(1 + k\frac{\mu B_{eq}}{E}\right)\right)\right]\right].$$

We give the following integral

$$\int_0^{E/B} \frac{d\mu}{|v_\||} \mu^n = \sqrt{2m}\frac{(2n)!!}{(2n+1)!!}\frac{E^{n+1/2}}{B^{n+1}}, \tag{167}$$

and we perform the μ integration (remember that for the 2D dipole $\overline{\omega}_{di}$ is independent of μ):

$$N = \int \frac{dl}{B}\left[(\frac{q_i}{m_i})^2 \int 4\pi dE f_{0i}\sqrt{2m_i}E^{1/2}\left[\frac{\overline{\omega}_{di}(\omega_{*i} - \overline{\omega}_{di})}{\omega(\omega + \overline{\omega}_{di})}\lambda_{neq}\left(c_n\right.\right.\right.$$

$$+ 2\sum_{k=0}^{n+1}(-1)^k C_{n+1}^k \frac{(2k-1)!!}{(2k+2)!!}\sum_{j=0}^{k}(-1)^j C_k^j \left(\frac{B_{eq}}{B}\right)^j \frac{(2j)!!}{(2j+1)!!}$$

$$\left.\left.\left.\left(1 + k\frac{(2j+2)}{(2j+3)}\frac{B_{eq}}{B}\right)\right)\right]\right].$$

The denominator D of 166 gives:

$$\int \frac{dl}{B}\left[(\frac{q_i}{m_i})^2 \int \frac{4\pi B dE d\mu}{|v_\||} f_{0i} \times \left[\frac{\overline{\omega}_{di}(\overline{\omega}_{di} - \omega_{*i})}{\omega(\omega + \overline{\omega}_{di})}\right]\right] =$$

$$\int \frac{dl}{B}\left[(\frac{q_i}{m_i})^2 \int 4\pi dE f_{0i}\sqrt{2m_i}E^{1/2} \times \left[\frac{\overline{\omega}_{di}(\overline{\omega}_{di} - \omega_{*i})}{\omega(\omega + \overline{\omega}_{di})}\right]\right]. \tag{168}$$

The terms being integrated over the energy vanish between the numerator N and the denominator D, we obtain:

$$\Phi_0 = \int \frac{dl}{B}\left[\lambda_{neq}\left(c_n + 2\sum_{k=0}^{n+1}(-1)^k C_{n+1}^k \frac{(2k-1)!!}{(2k+2)!!}\right.\right.$$

$$\left.\left. \times \sum_{j=0}^{k}(-1)^j C_k^j \left(\frac{B_{eq}}{B}\right)^j \frac{(2j)!!}{(2j+1)!!}\left(1 + k\frac{(2j+2)}{(2j+3)}\frac{B_{eq}}{B}\right)\right)\right] / \int \frac{dl}{B}.$$

Remember that $B = B_{eq}/\sin^2\theta$, we perform an l integration and we obtain:

$$\Phi_0 = \int \frac{dl}{B}\left[\lambda_{neq}\left(c_n + \frac{4}{\pi}\sum_{k=0}^{n+1}(-1)^k C_{n+1}^k \frac{(2k-1)!!}{(2k+2)!!}\right.\right.$$

$$\sum_{j=0}^{k}(-1)^j C_k^j \frac{(2j)!!}{(2j+1)!!}\pi^{1/2}\frac{\Gamma(j+3/2)}{\Gamma(j+2)}$$

$$\left.\left.\left(1+k\frac{(2j+2)(j+3/2)}{(2j+3)(j+2)}\right)\right)\right]/\int\frac{dl}{B}.$$

where we have used the following integral

$$\int_0^\pi d\theta \sin^p\theta = \pi^{1/2}\frac{\Gamma((p+1)/2)}{\Gamma(p/2+1)}, \tag{169}$$

where Γ is the classical Gamma function ($\Gamma(n+1/2) = (2n-1)!!\pi^{1/2}/2^n$). Finally, we obtain:

$$\Phi_0 = \lambda_{neq}\left(c_n + S_n\right), \tag{170}$$

where we have defined

$$S_n = \sum_{k=0}^{n+1}(-1)^k C_{n+1}^k \frac{(2k-1)!!}{(2k+2)!!}\sum_{j=0}^{k}(-1)^j C_k^j \frac{(2j)!!}{2^{j-1}(j+1)!}\left(1+k\frac{j+1}{j+2}\right). \tag{171}$$

E   Calculation of the Bounce Averaged $(\delta E_y \times B)/B^2$ Drift

Taking into account the expression (51) of the perpendicular electric field δE_y, the bounce averaged electric drift writes:

$$\overline{u_{E_y}} = \frac{1}{\tau_b}\oint\frac{dl}{|v_\parallel|}\frac{\delta E_y}{B}$$

$$= -\frac{4L}{v\tau_b}\frac{\delta E_{L,y,t}(L,y,t)}{B_{eq}}\frac{1}{n+1}$$

$$\times\int_{\frac{\pi}{2}}^{\theta_m}\frac{d\theta}{\left(1-\frac{\mu B_{eq}}{E\sin^2\theta}\right)^{1/2}}\left(S_n - (\sin^2\theta)^{n+1}\right)\sin^2\theta.$$

Using the same change of variables as in Appendix C, we obtain

$$\overline{u_{E_y}} = \frac{4L}{v\tau_b}\frac{\delta E_{L,y,t}(L,y,t)}{B_{eq}}\frac{1}{n+1}$$

$$\times\int_0^1\frac{dZ}{(1-Z^2)}\left(S_n(1-AZ^2)-(1-AZ^2)^{n+2}\right). \tag{172}$$

Again, we use the binomial formula and perform the Z integration (see Appendix C), we get:

$$\overline{u_{E_y}} = \frac{4L}{v\tau_b} \frac{\delta E_{L,y,t}(L,y,t)}{B_{eq}} \frac{1}{n+1} \left(S_n \frac{\pi}{2} (1 - \frac{A}{2}) \right.$$
$$\left. - \sum_{k=0}^{n+2} (-1)^k C_{n+2}^k A^k \frac{(2k-1)!!}{(2k+2)!!} \frac{\pi}{2} \right). \quad (173)$$

Then, we substitute the expression of $\tau_b = 2\pi L/v$ and $A = (1 - \mu B_{eq}/E)$ and obtain

$$\overline{u_{E_y}} = \frac{\delta E_{L,y,t}(L,y,t)}{B_{eq}} \frac{1}{n+1} \left[\frac{S_n}{2} \left(1 + \frac{\mu B_{eq}}{E} \right) \right.$$
$$\left. - \sum_{k=0}^{n+2} (-1)^k C_{n+2}^k \frac{(2k-1)!!}{(2k+2)!!} \sum_{j=0}^{k} (-1)^j C_k^j \left(\frac{\mu B_{eq}}{E} \right)^j \right]. \quad (174)$$

References

[1] Sagdeev R.Z., Galeev A.A. (1969) Non-linear Plasma Theory. T. M. O'Neil and D. L. Brooks

[2] Biskamp D. (1986) Details of current disruption and diversion in simulations of magnetotail dynamics. Phys. Fluids **29**, 1520

[3] Chen L., Hasegawa A. (1988) On magnetospheric hydromagnetic waves excited by energetic ring-current particles. J. Geophys. Res. **93**, 8763

[4] Chen L., Hasegawa A. (1991) Kinetic theory of geomagnetic pulsations, 1. internal excitations by energetic particles. J. Geophys. Res. **96**, 1503

[5] Cheng C.Z. (1982) Kinetic theory of collisionless ballooning modes. Phys. Fluids **25**, 1020

[6] Cheng C.Z., Johnson J.R. (1999) A kinetic-fluid model. J. Geophys. Res. **104**, 413

[7] Démoulin P., Klein K.L. (1998) Structuring of the solar plasma by the magnetic field. this volume

[8] Somov B.V. (1992) Solar Flares. Spinger, Berlin

[9] Pellat R., Hurricane O.A., Luciani J.F. (1996) New mechanism for three-dimensional current dissipation/reconnection in astrophysical plasmas. Phys. Pev. Lett. **77**, 4354

[10] McIlwain C.E. (1975) Auroral electron beams near the magnetic equator. In: Hultqvist B., Stenflo L. (eds.), *Physics of the hot plasma in the magnetosphere*. Plenum Pub. Corp., New York

[11] Parks G.K., Lin C.S., Mauk B., DeForest S., McIlwain C.E. (1977) Characteristics of magnetospheric particle injection deduced from events observed on august 18, 1974. J. Geophys. Res. **82**, 5208

[12] Matsushita K., Masuda S., Kosugi T., Inda M., Yaji K. (1992) The average height of hard x-ray sources in solar flares. Pub. Astron. Soc. of Japan **44**, L89

[13] Coppi B., Laval G., Pellat R. (1966) Dynamics of the geomagnetic tail. Phys. Rev. Lett. **16**, 1207

[14] Pellat R., Coroniti F.V., Pritchett P.L. (1991) Does ion tearing exist? Geophys. res. Lett. **18**, 143

[15] Pellat R. (1990) Une nouvelle approche de la reconnexion magnétique : Sous-orages magnétosphériques - vents stellaires. C. R. Acad. Sci. **311**, 1706

[16] Ichimaru S. (1973) Basic Principles of Plasma Physics. W. A. Benjamin, Inc, Massachusetts

[17] Rutherford P.H., Frieman E.A. (1968) Drift instabilities in general magnetic field configurations. Phys. Fluids **11**, 569

[18] Hurricane O.A. (1994) The kinetic theory and stability of a stochastic plasma with respect to low frequency perturbations and magnetospheric convection. Ph.D. thesis, Université de Californie, Los Angeles

[19] Hurricane O.A., Pellat R., Coroniti F.V. (1995) The stability of a stochastic plasma with respect to low frequency perturbations. Phys. Plasmas **2**, 289

[20] Hurricane O.A., Pellat R., Coroniti F.V. (1995) A new approach to low-frequency "mhd-like" waves in magnetospheric plasmas. J. Geophys. Res. **100**, 19421

[21] Antonsen Jr. T.M., Lane B. (1980) Kinetic equations for low frequency instabilities in inhomogeneous plasmas. Phys. Fluids **23**, 1205

[22] Lui A.T., Lopez R.E., Anderson B.J., Takahashi K., Zanetti L.J., McEntire R.W., Potemra T.A., Klumpar D.M., Greene E.M., Strangeway R. (1992) Current disruption in the near-earth neutral sheet region. J. Geophys. Res. **97**, 1461

[23] Baumjohann W., Paschmann G., Cattell C.A. (1989) Average plasma properties in the central plasma sheet. J. Geophys. Res. **94**, 6597

[24] Slavin J.A., Smith E.J., Sibeck D.G., Baker D.N., Zwickl R.D., Akasofu S.I. (1965) An isee 3 study of average and substorm conditions in the distant tail. J. Geophys. Res. **90**, 10875

[25] Frank L.A., Paterson W.R., Ackerson K.L., Kokubun S., Yamamoto T. (1996) Plasma velocity distributions in the near-earth plasma sheet: A first look with the geotail spacecraft. J. Geophys. Res. **101**, 10627

[26] Pellat R., Hurricane O.A., Coroniti F.V. (1994) Multipole stability revisited. Phys. Plasmas **1**, 3502

[27] Bernstein I.B., Frieman E.A., Kruskal M.D., Kulsrud R.M. (1958) An energy principle for hydromagnetic stability. Proc. Roy. Soc. (London) **A 244**, 17

[28] Hurricane O.A., Pellat R., Coroniti F.V. (1994) The kinetic reponse of a stochastic plasma to low frequency perturbations. Geophys. Res. Let. **21**, 253

[29] Jacquey C. (1996) Time-variation of the large-scale tail magnetic field prior to substorm related to solar wind changes. In: *Third International Conference on Substorms, Versailles, 12-17 mai, 1996*

[30] Huang T.S., Birmingham T.J. (1994) Kinetic and thermodynamic properties of a convecting plasma in a two dimensional dipole field. J. Geophys. Res. **99**, 17295

[31] Sauvaud J.A., Winckler J.R. (1980) Dynamics of plasma, energetic particles, and fields near synchronous orbit in the nighttime sector during magnetospheric substorms. J. Geophys. Res. **85**, 2043

[32] Sergeev V.A., Mitchell D.G., Russel C.T., Williams D.J. (1993) Structure of the tail plasma/current sheet at $\simeq 11r_e$ and its changes in the course of a substorm. J. Geophys. Res. **98**, 17345

[33] McPherron R.L. (1979) Magnetospheric substorms. Rev. Geophys. **17**, 657

[34] Roux A., Perraut S., Robert P., Morane A., Pedersen A., Korth A., Kremser G., Aparicio B., Rodgers D., Pellinen R. (1991) Plasma sheet instability related to the westward traveling surge. J. Geophys. Res. **96**, 17697

[35] Nagai T. (1991) An empirical model of substorm-related magnetic field variations at synchronous orbit. In: Kan J.R., Potemra T.R., Kokobun S., Iijima T. (eds.), *Magnetospheric Substorms (Proc. Chapman Conf., Hakone Japan, Sept. 3-7, 1990)*. Washington, DC

[36] Korth A., Pu Z.Y., Kremser G., Roux A. (1991) A statistical study of substorm onset conditions at geostationnary orbit. In: Kan J.R., Potemra T.R., Kokobun S., Iijima T. (eds.), *Magnetospheric Substorms (Proc. Chapman Conf., Hakone Japan, Sept. 3-7, 1990)*. Washington, DC

[37] Perraut S., Morane A., Roux A., Pedersen A., Schmidt R., Korth A., Kremser G., Aparicio B., Pellinen R. (1993) Characterization of small scale turbulence observed at substorm onsets: relationship with parallel acceleration of particles. Adv. Space Res. **13**, 217

[38] Perraut S., Roux A., Le Contel O., Pellat R., Pedersen A., Korth A. (1998) Evidence for a substorm trigger. In: *Fourth International Conference on Substorms (ICS-4), Lake Hamana, Japan, March 9-13, 1998*

[39] Koskinen H.E.J., Toivanen P.K., Pulkkinen T.I. (1996) Parallel electric fields during the substorm growth phase. In: *Third International Conference on Substorms, Versailles, 12-17 mai, 1996*

[40] Zwillinger D. (1989) Handbook of differential equations. Harcourt Brace Jovanovich, Boston

Lecture Notes in Physics

For information about Vols. 1–519
please contact your bookseller or Springer-Verlag

Monographs
For information about Vols. 1–20
please contact your bookseller or Springer-Verlag